ROOF CONSTRUCTION AND REPAIR

By E Molloy

With an introduction by Richard Jordan and Tim Ratcliffe

Routledge
Taylor & Francis Group
LONDON AND NEW YORK

First edition published in 1941 by George Newnes Limited, London.
Second edition published in 1945 by George Newnes Limited.

Published by Donhead Publishing Ltd 2009

Published 2015 by Routledge
2 Park Square, Milton Park, Abingdon, Oxon OX14 4RN
711 Third Avenue, New York, NY 10017, USA

Routledge is an imprint of the Taylor & Francis Group, an informa business

ISBN 13: 978-1-873394-88-5 (hbk)

British Library Cataloguing in Publication Data

Molloy, E.
Roof construction and repair.
1. Roofs--Design and construction. 2. Tiles, Roofing.
I. Title
690.1'5-dc22
ISBN-13: 9781873394885

Introduction to the 2009 edition

Roof Construction and Repair was first published in 1942, during the Second World War, and republished in a second slightly enlarged edition in 1945. The publication of new books was limited in wartime, and the fact that this was considered 'worth the paper' illustrates the practical concerns of the period. As so many skilled roofers had been sent off to fight in the war, the book was needed to give its readers a clear and practical understanding of roofing, even with limited prior knowledge.

Although the text contains simple explanations, and 'step by step' guidance, it is clear that Mr Molloy is a craftsman with considerable knowledge and a real passion for his subject. His real skill is in communicating the complexities of good roofing in ways that are easy to understand. This is the reason his writing is so useful for today's reader.

Anyone stimulated to absorb the level and depth of knowledge of a traditional roofer, who learned his trade before the last world war, will benefit from reading this book. About three quarters of the content relates to traditional roofing materials and techniques, and will be of direct interest and use to those involved with repairing old buildings. The remaining quarter describes materials and techniques that were prevalent at the time. Although these sections may not seem so immediately relevant, the fact that asbestos was used in some roofing felts during this period, as well as in cement tiles, is worth highlighting.

It is no surprise that there is a whole chapter on patching roofs and emergency repair methods. Although the concern was with damage caused by air raids, the methods advocated are relevant to anyone trying to patch a roof to prevent water ingress. Indeed, some of the suggestions highlight the possibilities for repairing roofs, rather than stripping and re-laying them. This is an issue that deserves further thought, as roof coverings are sometimes replaced before it is actually necessary.

This reprint of *Roof Construction and Repair* is of the second edition. This includes additional information on the use of asphalt shingles and also new materials for flat roofs plus, more significantly, further diagrams and illustrations.

The first four chapters of the book describe the timber construction of roofs, covering everything from basic joints through to the differences in the design of traditional and trussed roofs. There is a substantial amount of practical information on the design of pitched and flat roofs, the design of jointing around roof lanterns, fixing lay boards up the pitch of the roofs and the setting out of steps in horizontal gutters. The calculations in Chapter IV on the sizing of roof timbers are quite complicated, and are partially superseded by the 'Roof Tables' in the last chapter, added into the second edition.

Chapter V describes slates and slating, and is illustrated with some good photos. The examples are generally from Delabole and have a West Country emphasis. It explains the principles relating to the 'coverage' of slates, and how to calculate the quantities required. It also includes descriptions of how to sort and grade slates, how to work out of the position of the nail holes in slates and then how to hole them. The chapter ends with random slating and notes 'the gradual decrease in the gauge to give a 3-inch lap throughout should be noted'. This should really emphasize that a *minimum* lap of 3-inches is required.

Chapter VI explores the tiling of roofs, and begins with a very useful description of the different types and variations of tile used historically. This includes an explanation of the ancient Roman method of alternating over and under tiles (Imbrex and Tegula), which developed into the profiled single lap tiles that we still refer to as 'Roman', and also into 'Spanish' tiling. There is then a clear explanation of the setting out of the gauge and lap of a plain tiled roof, including the difference between rafter pitch and effective pitch, followed by a description of details such as verges, ridges, abutments and valleys. The final section covers single-lap interlocking tiles and different ways to detail hips and valleys.

The next three chapters discuss roofing with asbestos-cement based materials, asphalt shingles, and asbestos felt. Although most of these materials are not used today, some, like diamond slating and fibre cement slating, are being reused or repaired in conservation areas. We still need to learn about these products, as it is not only important to understand how to repair and possibly re-lay them, but also to be aware of the risk of asbestos and consequent health issues.

Chapter X describes the use of wood shingles, which was increasing in popularity at the time. This section is rather basic, but is a useful introduction. Unfortunately it doesn't differentiate between the durability of cedar and oak, or between sawn and split shingles. We now understand that oak shingles, particularly if they have been split (to follow the grain), will last considerably longer than cedar.

Emergency repairs to roofs after an air raid are described in Chapter XI. It offers useful advice on how to keep a building watertight until financial resources or materials can be found to re-cover the roof properly (a concern these days when trying to save a redundant building from further decay). This includes guidance on laying felt or corrugated iron over a roof, and the best ways to finish edges. There is also a suggestion about how to lay slates in 'extended order' (i.e. with vertical gaps) if there is a shortage of materials. This is advocated as a temporary measure, but there are examples of roofs laid this way, to be found today, that are still serviceable. Slightly worryingly, the text also makes reference to 'slurrying' to avoid damaging undisturbed areas of a roof. Presumably this refers to the application of a cement slurry or bitumen (the latter known as 'turnerising'), both of which cause damage to slates and tiles in the long term.

The final chapter is titled 'Roofing Tables'. This gives information on the sizes of timbers required for different rafter lengths on pitched roofs, in tabular form.

This book, with its simple but comprehensive description of roofing methods, is likely to have saved many damaged buildings from further decay during the war. As well as the chapter on emergency repairs, the other chapters give enough information for a practical person to lay a slate or tile roof properly. Tim Ratcliffe has photos of his grandfather, a vicar in Dagenham during the war, on the roof of the church, refixing slates after a bombing raid. He seems to be doing a reasonable job and it would be nice to think that he, and the curate with him, had a copy of *Roof Construction and Repair* to refer to!

We should be clear, however, that this is not just a beginner's guide to roofing. Mr Molloy's knowledge means that the level of information and detail goes far beyond this. This is a book that deserves to be more widely known and will prove useful to anyone concerned with the skills and craft of roofing.

Richard Jordan & Tim Ratcliffe
August 2009

Richard Jordan (Roofer)
Richard Jordan is a slater who followed in his father's footsteps, and learnt his trade working with him on site. In 1999 he was awarded a 'William Morris' travelling fellowship by the Society for the Protection of Ancient Buildings (SPAB). He has worked on the repair and replacement of roofs on historic buildings for a variety of clients, including the National Trust. He now also provides specialist consultancy advice on roofs and roofing projects, and is involved with encouraging training in traditional craft skills for building owners, architects, contractors and other construction professionals.

Tim Ratcliffe (Architect)
Tim Ratcliffe was awarded a 'Lethaby' travelling scholarship by the SPAB in 1987 and has also worked as a labourer for a number of specialist conservation contractors. He worked for two architectural firms specialising in historic building work (Rodney Melville & Partners and Donald Insall Associates) before setting up in practice with his wife, Jan, in 2000. Based in Oswestry, they work and give advice on churches, houses, castles and old industrial buildings in the Midlands and North Wales.

ROOF CONSTRUCTION AND REPAIR

ROOF CONSTRUCTION AND REPAIR

DEALING WITH SLATE, TILE, ASBESTOS-CEMENT, FELT, AND CONCRETE ROOFS, WITH A SPECIAL CHAPTER ON EMERGENCY REPAIRS

Prepared by a Staff of Technical Experts under the direction of

E. MOLLOY

WITH ONE HUNDRED AND EIGHTY-ONE ILLUSTRATIONS

GEORGE NEWNES LIMITED

TOWER HOUSE, SOUTHAMPTON STREET, STRAND

LONDON, W.C.2

First Published . . . 1942
Second Edition . . . 1945

PRINTED IN GREAT BRITAIN BY HAZELL, WATSON & VINEY, LTD., LONDON AND AYLESBURY

PREFACE

ROOF construction may be divided broadly into two sections:

(*a*) carpentry work, which is concerned with the construction of roof framework, and

(*b*) roof covering, which involves the application of slates, tiles, asbestos cement, and similar materials, to the roof structure. The present book deals with both aspects of the subject.

In the first chapter will be found a clear explanation of the various components which together form the roof structure. All the main types, including the collar-and-tie roof, the valley roof, the purlin roof, and the Mansard roof are described and illustrated.

The second chapter is devoted to the construction of trussed roofs, which are used in most cases where the span exceeds about 20 ft.

Flat roofs constructed of either timber or reinforced concrete form the subject of Chapter III, and in this section will be found, in addition to the structural details, useful information concerning the chief types of coverings which are employed for flat roofs, namely, asphalte, copper, and zinc. An item of particular interest which is also dealt with in this chapter is the subject of roof lights employing "Glass-crete," i.e. a series of glass prisms embedded in concrete. Another important feature which is closely allied to the subject of flat roofs is that of heat insulation. This also receives adequate treatment in the section under review.

The various calculations involved in determining the sizes and spacing of the various roof members form the subject of the next chapter.

In spite of the growing popularity of natural tiles and also composition tiles, the slate roof still holds pride of place. All the practical details which are necessary in connection with roof slating are given in Chapter V, a particularly valuable feature being the trade descriptions of the various sizes and types of roofing slates, which range from Princesses, size 24 in. by 14 in., down to Small Doubles, size 12 in. by 6 in. In this chapter also will be found some most useful information enabling the quantities of slates required for a given job to be accurately estimated.

After these preliminary details, practical instructions are given for the actual operation of slating. Similar information is also given concerning all the important forms of tiles, both interlocking and plain.

Asbestos-cement tiles, under which classification are included a number of well-known proprietary makes, are dealt with in Chapter VII. Although some of these are similar in appearance and dimensions to the natural clay or slate product, the methods of fixing vary considerably. All these variations have been carefully explained from the point of view of the practical roofer and tiler.

The method of applying asbestos-felt, and the use of wood shingles as a roof covering, form the subjects of the succeeding chapters. Emergency repairs to roofs have been dealt with in considerable detail in the final chapter. Preliminary inspection, emergency slating repairs, covering roofs with bitumen roofing felt, the use of corrugated iron and ruberoid roofing materials are all carefully explained in detail with a wealth of illustration.

We should like to take this opportunity of thanking the suppliers of roofing materials for the valuable assistance which they have rendered us in the preparation of this work. In particular we would express our indebtedness to the Westmorland Green Slate Quarries Ltd., Messrs. Langley London Ltd., The Marley Tile Co., Messrs. Turners Asbestos Co. Ltd., The Universal Asbestos Manufacturing Co. Ltd., The Atlas Stone Co. Ltd., W H. Colt Ltd., Messrs. D. Anderson & Sons Ltd., and Messrs. Frazzi Ltd. for supplying technical data and trade information concerning their respective products.

It is hoped that the collection of this information into one volume, convenient for reference, will prove of maximum utility for all concerned in the design and construction and repair of roofs.

E. M.

PREFACE TO SECOND EDITION

In revising this book the opportunity has been taken of adding a new chapter on Covering Roofs with Asphalte Shingles. New material has also been included relating to materials for covering flat roofs, together with many useful diagrams.

E. M.

CONTENTS

PAGE

PREFACE v

CHAPTER I

ROOF CONSTRUCTION FOR SMALL BUILDINGS 1

Types of Roofs—Span—Pitch of Roof—Parts of a Roof—Single Roofs—Couple Roofs—Couple-close Roof—Collar-tie Roof—Valley or Double Lean-to Roof—Purlins—Roofs Constructed of Short Timbers—Single Mansard Roof—Eaves Finish for Roofs—Precast Concrete Roofing Units—Trimming.

CHAPTER II

TRUSSED ROOFS 18

King-post Truss—Sizes of Various Members—Mansard or Curb Roof—Composite Roof Trusses.

CHAPTER III

FLAT ROOFS 28

Timber Construction—Lead Flats, Drips—Rolls for Lead—Zinc Roofing—Copper Roofing—Asphalte Roofing—Steel Construction—" Glass-crete "—Roof Insulation—Thermotile Roofing—Paropa—Parallel Gutters—Tapering Gutters.

CHAPTER IV

SIZES AND SPACING OF ROOF MEMBERS 43

Working Stresses for Rafters—Resistance Moment—Design of Boarding—Rafter Design—Design of Purlins—Stress Diagrams for Roof Trusses—Roof Bevels.

CHAPTER V

SLATES AND SLATING 58

Quality, Weight, and Thickness—Calculations for Quantities—Sized Slates—How Slates are Sold—Random Slates—Delabole Randoms—Calculations for Battening—Preparation for Slating—Eaves—Verges—Ridges—Hips—Valleys—Abutments—Nailing—Roof Pitches.

CHAPTER VI

HOW TO TILE ROOFS 70

Clay Tiles—Qualities—Classification of Clay Tiles—Old English Pantile—Double Roman—Single Roman—Double Roman Interlocking—Somerset Interlocking No. 13—Reynardo—Spanish—Italian—Courtrai-du-Nord—Interlocking Pantile—Cloister—Plain—Calculations—Plain Tiling—Ridge Tiles—Bonnet Hips, Hip Tiles, and Valley Tiles—Angle Tiles—Single-lap Tiling—Weights—Choosing a Tile—Special Tiles—Concrete Tiles.

PAGE

CHAPTER VII

ROOFING WITH ASBESTOS-CEMENT MATERIALS 96

Roofing Materials—Asbestos-cement Slates—Corrugated Sheets—Pantiles—Bedding.

CHAPTER VIII

COVERING ROOFS WITH ASPHALTE SHINGLES 103

Laying of Flextile Shingles—Lap—Nailing—Edges and Hips—Flashings.

CHAPTER IX

COVERING ROOFS WITH ASBESTOS-FELT 107

Standard Weights and Measurements of Material—Valley Gutters—Lap Cement—Clout Nails—Flashings—Sarking.

CHAPTER X

COVERING ROOFS WITH WOOD SHINGLES 110

Types—Size—Weight—Nailing—Laying—Hips and Ridges—Valleys—Walls—Colour.

CHAPTER XI

EMERGENCY REPAIRS TO ROOFS 113

Roof Inspection—Slate-covering Repair—Tile-covering Repair—Covering Roofs with Bitumen Roofing Felt—Fitting Corrugated-iron Roofing—Fixing Ruberoid Roofing.

ROOFING TABLES 126

INDEX 134

ROOF CONSTRUCTION AND REPAIR

Chapter I

ROOF CONSTRUCTION FOR SMALL BUILDINGS

ROOF construction is an interesting subject at any time to all engaged in the work of house building and repair. At the present time, however, the subject is one of vital importance owing to the enormous amount of damage which has occurred to roofs as a result of enemy air activity. When such damage occurs, the subject of speedy repair is of the utmost importance if the building is to be rendered habitable.

Before roof work can be undertaken, it is essential to understand the various forms of roof construction that are to be met with. This aspect receives attention in the first four chapters of the present work. Special attention is given to the methods employed for domestic dwellings in which timber construction is largely employed. The remaining chapters deal with the materials used for coverings, and how these coverings are placed on the roof members. This section includes a chapter on the emergency repair of roofs.

Types of Roofs

There are several types of roofs, which are designated by various names according to outline or form of construction. Many considerations are taken into account when determining the type of roof to be adopted, such as span, material used for the covering, and climatic conditions.

Span

The *effective* span of a roof is the horizontal distance between the centres of the bearings on the supports, and the *clear* span the horizontal distance between the walls.

Pitch of Roof for Various Coverings

The term *pitch* is applied to the amount of slope given to the side of a roof, and may be stated in terms of the number of degrees in the angle which the slope makes with the horizontal, or the ratio of the rise to the span which is usually designated by a fraction. Thus, suppose the rise to be 12 ft. and the span 48 ft., then the pitch is :

$$\frac{\text{rise}}{\text{span}} = \frac{12}{48} = \frac{1}{4}$$

The minimum pitch or inclination it is desirable to use for the various materials used for coverings is given in the table on the next page.

PITCH OF ROOF FOR VARIOUS MATERIALS

Materials	*Pitch*	
	Minimum Angle with Horizon	*Ratio of Rise to Span*
Slates (small)	33°	$\frac{1}{3}$
,, (ordinary)	26°	$\frac{1}{4}$
,, (large)	21°	$\frac{1}{5}$
Tiles (single-lap)	35°	$\frac{7}{20}$
Plain tiles	45°	$\frac{1}{2}$
Corrugated sheets	20°	$\frac{9}{50}$
Copper and zinc	1°	$\frac{1}{80}$
Lead and asphalte	0°	$\frac{1}{120}$
Ruberoid	20°	$\frac{9}{50}$

In exposed positions these pitches should be increased over the amounts given above.

Parts of a Roof

The lowest part or edge of a sloping roof is called the *eaves,* and may or may not project beyond the wall face. Where the roof finishes over a gable end, the edge is termed the *verge. Valley* is the line of intersection of two roof planes containing an external angle less than 180°, and a *hip* the line of intersection of two roof surfaces which contain an angle greater than 180° A *hipped end,* the triangular surface formed by the slope being continued round the end. These terms are illustrated in Fig. 1. *Rafters* (common) are inclined members running from ridge to eaves. Where they finish against hip or valley rafters they are termed *jack rafters.*

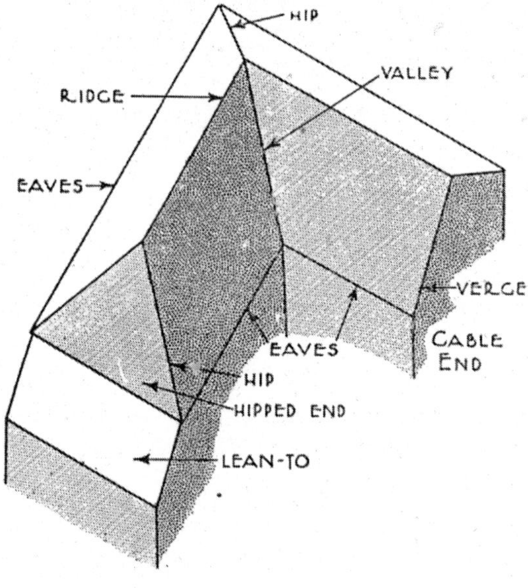

Fig. 1.—SKETCH OF ROOF, ILLUSTRATING THE TERMS EMPLOYED TO DESCRIBE PARTS

A series of rafters 12 in. to 16 in. apart is usually placed over the entire surface. If they exceed about 8 ft. in length, intermediate supports termed *purlins* must be provided for them. Generally, they run from the roof ridge or apex down to

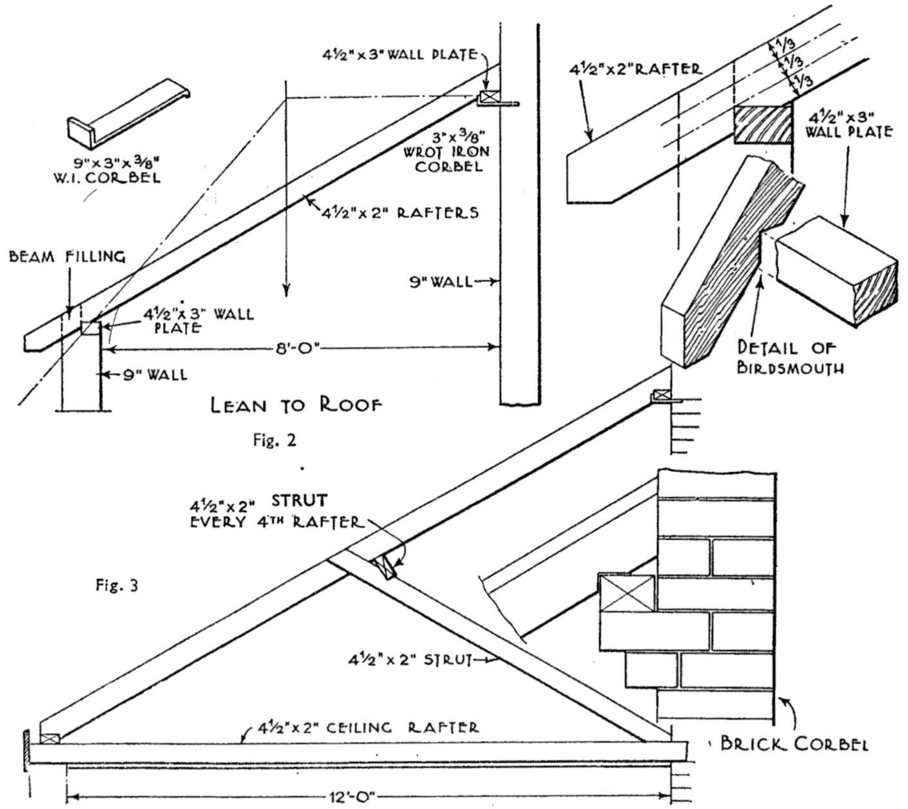

Figs. 2 *and* 3.—LEAN-TO ROOFS

The span suitable for the type shown in Fig. 2 is 8 ft., but it can be increased if suitably tied and braced by the use of purlins and struts at intervals if necessary, as in Fig. 3. Fig. 3 also shows the construction of a brick corbel.

the eaves, but occasionally are placed in a horizontal position. *Wall plates*, to which the lower ends of the rafters are usually fitted with a bird's-mouth joint and nailed, should be bedded on the wall, preferably placed along the inside. For convenience of fixing the roof covering, it is necessary either to nail battens transversely at intervals or to board the rafters entirely over.

Roofs may be classified as single, trussed, composite (wood and steel), and steel.

Single Roofs

Single roofs are those which consist of a framework of common rafters alone, with a wall plate and a ridge piece, and with or without purlins. These include the following types: lean-to, couple, couple-close, collar-tie,

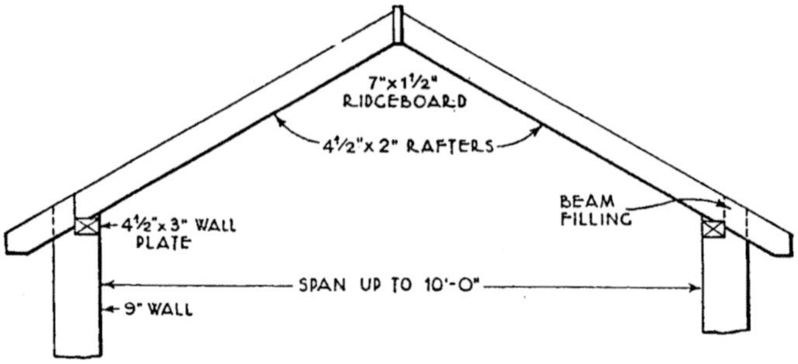

Fig. 4.—Couple roof

The weak feature of this type is that its weight does not bear vertically on the walls.

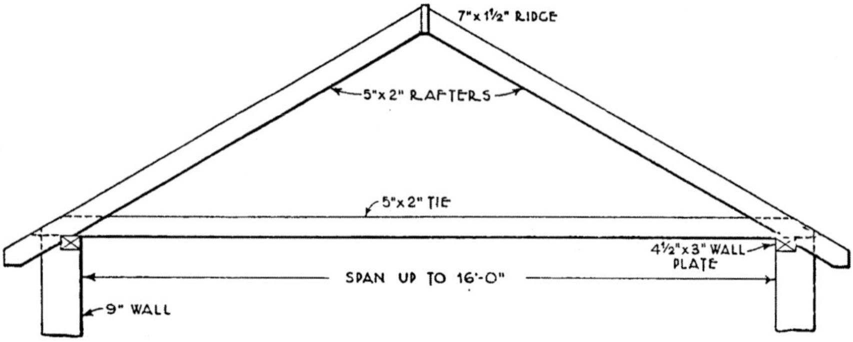

Fig. 5.—Couple-close roof

The couple-close roof is tied at the feet of the rafters. The ties act as ceiling joists where a ceiling is required.

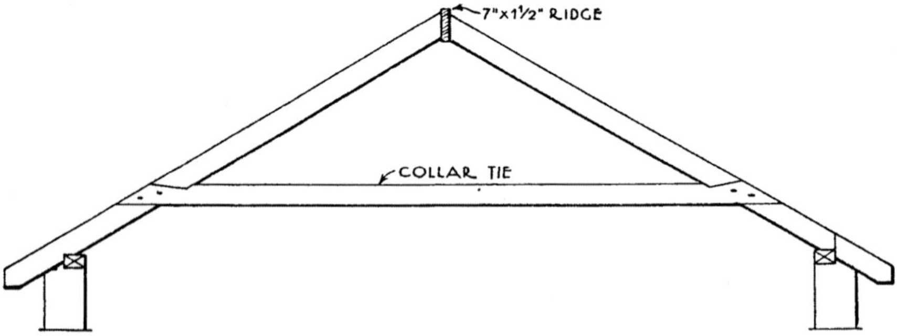

Fig. 6.—Collar-tie roof

The tie is placed some distance above the wall plate, and is called a tie-rod if of steel and a tie-beam if of wood. It is usual to place the tie at one-third the vertical height from wall to ridge.

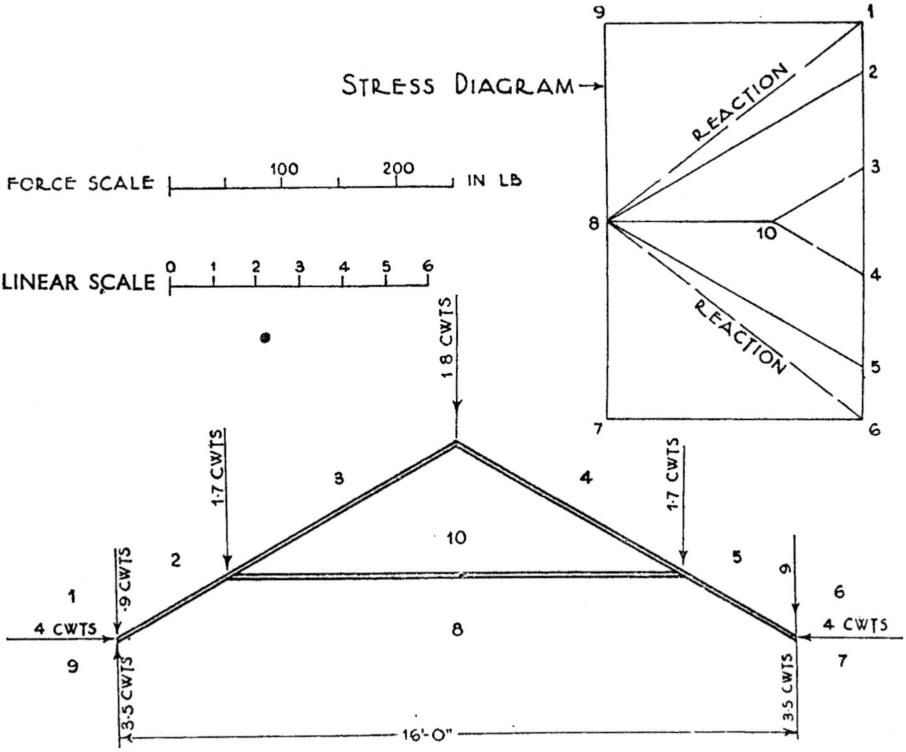

Fig. 7.—FRAME DIAGRAM FOR RIGID WALLS

V-roof, and flat roof. When purlins are used they are sometimes called double or purlin roofs.

Lean-to

This is formed with only one slope, and is used for sheds and buildings attached to the main buildings (see Fig. 2). The span suitable for this type is 8 ft., but it can be increased if suitably tied and braced by the use of purlins and struts at intervals if necessary (see Fig. 3). If the distance from ridge to eaves is less than 5 ft., it is not necessary to use rafters at all, but simply to board the surface over, the boards running with the slope, and a suitable support provided for them at top and bottom.

The upper ends of the rafters are fixed to a wall plate which is either built into or attached to the back wall at the desired height. When the wall plate is not built into the wall, or rests on an offset, it may be supported on a projecting course of bricks, or corbels of brick or wrought iron, at about 3-ft. intervals, built in or fixed for the purpose (see Figs. 2 and 3), The lower ends of the rafters are fitted to the wall plate in the usual way

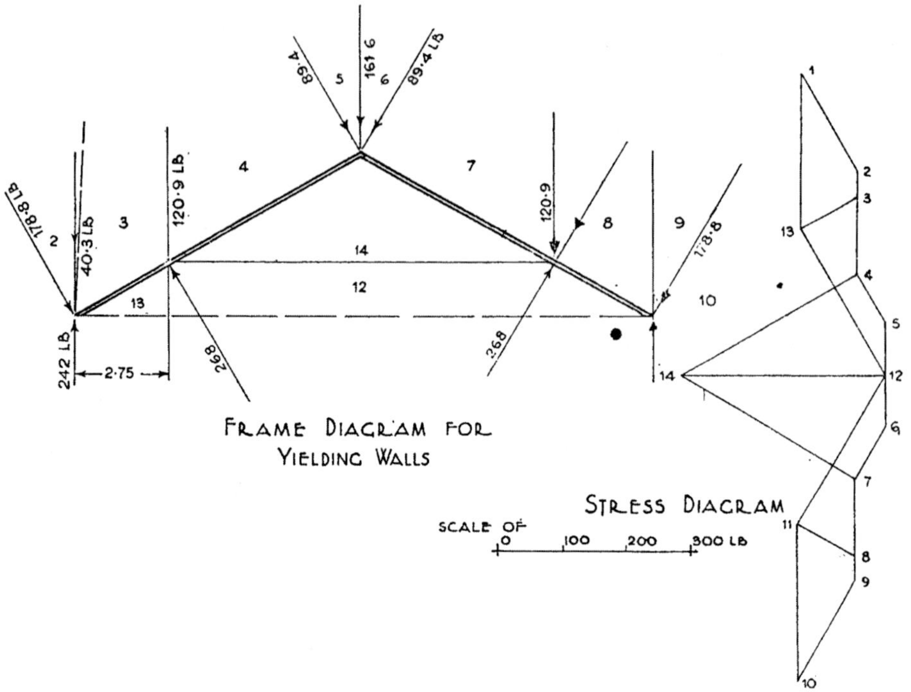

Fig. 8.—STRESSES IN COLLAR-TIE ROOF

Couple Roofs

As their name implies, couple roofs are roofs consisting of pairs of rafters fixed at their feet to a wall plate (see Fig. 4), and pitching against the ridge. The weak feature of these roofs is that their weight does not bear vertically on the walls, but tends to thrust them apart unless the walls are sufficiently strong to resist the thrust, but it is quite easy to modify its form so that all the outward thrust, or most of it, can be prevented. They are only suitable for spans up to about 10 ft.

Couple-close Roof

When the couple roof is tied at the feet of the rafters it is termed a *couple-close* roof (see Fig. 5). The weight of the roof, etc., tends to depress the ridge, and therefore the feet of the rafters will tend to spread and push the wall outwards. The tie prevents the feet from spreading, and consequently the danger of the walls overturning is avoided.

These ties also act as ceiling joists where a ceiling is required. Although the tie beam is in tension it requires to be of good depth to prevent sagging.

Where a ceiling is to be fixed a good rule is to make the depth of the joist ½ in. for every foot of span.

Collar-tie Roof

When the tie is placed some distance above the wall plate, it is called a *collar-tie* roof (see Fig. 6). This tie is called a tie-rod if of steel, and a tie-beam if of wood. If it is placed at half the vertical height from the wall plate to the ridge, the bending in the rafter would be prevented, but in this position the tie would not be very efficient to prevent the overturning thrust on the walls. It is therefore usual to compromise, and place it at one-third the vertical height from the wall to the ridge. The collars have a dual function, according to changing circumstances. If the walls are rigid and the rafters are sufficiently stiff to stand without deflection, the collar tie will be in compression (see Fig. 7), as also will be the case under wind pressure. However, the slightest degree of settlement or the spreading of the supports and they at once become ties (see Fig. 8). In view of this alternating change in function it is well that every fourth or fifth collar should be halved to the rafters in order to form a shoulder, but it is not wise to weaken the rafter at this point by cutting into it to form the joint. These rafters should be proportionately increased in width, so as to maintain the same sectional area, as the weak point of the roof is the amount of deflection that may occur in the rafter between the ends of the collars and the wall plate.

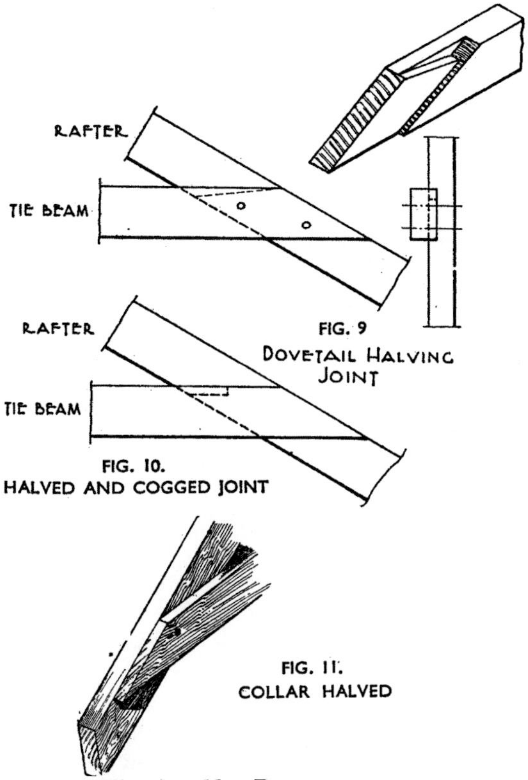

Figs. 9 *to* 11.—DETAILS OF JOINTS

The joint connecting the collar beam to the roof may be a dovetail-halved joint, as shown in Fig. 9, or a halved and cogged joint (Fig. 10), or the collar alone may be halved (Fig. 11).

Collar-and-tie Roof

Though the generality of textbooks deal almost entirely with roofs constructed with trusses, the greater number of small buildings are

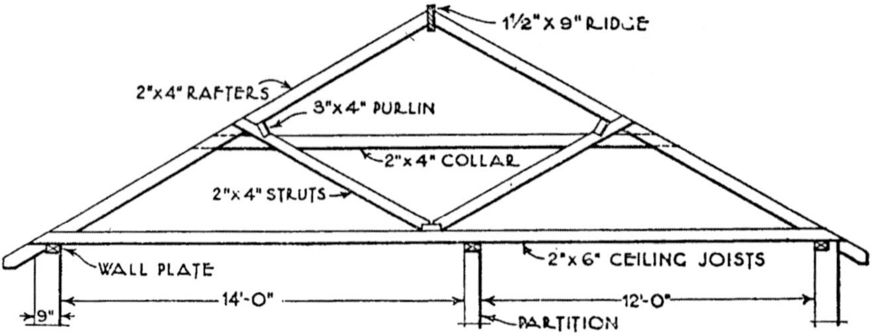

Fig. 12.—COLLAR-AND-TIE ROOF

without such provision. Besides the various forms of single roofs shown, larger spans can be satisfactorily constructed by the use of struts and ties. When constructed in this way they are known as *collar-and-tie* roofs (see Figs. 12 and 13). When this last type of roof is further strengthened by placing pairs of rafters at about 5 or 6 ft. apart, and securing with collars, struts, and hangers, they are termed *purlin* or *double-rafter* roofs. An alternative method is to carry the roof on a Howe or N girder, which forms a trussed purlin ; the top boom carries the rafters, and the lower one the ceiling rafters.

If the roof space is to be utilised as a room, then the collars will have to be raised to allow sufficient head room. The purlins can be used as ashlaring along the sides of the room and covered with a plaster board or fibre board, or the purlin filled in with studs to receive the laths for plastering.

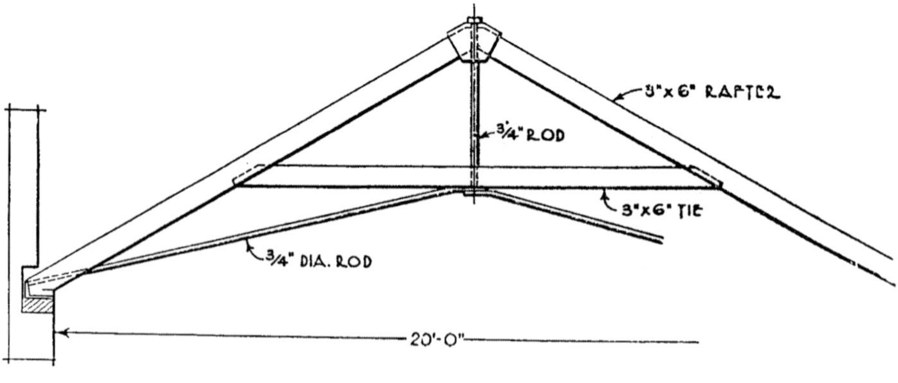

Fig. 13.—COLLAR-AND-TIE ROOF

Valley or Double Lean-to Roof

Another form of roof is the valley or double lean-to (see Fig. 14). The roof is inclined from the party or main walls, towards a gutter in the centre usually over a partition wall or some other support. The rafters rise from a pole plate, or sometimes from a pair of parallel plates which form the sides of the gutter, and which are supported by a partition. The upper ends of the rafters are carried on plates resting on wrought-iron or brick corbels from the party walls. The front of the building is usually finished with a wall and parapet, and the back of the building with gable ends, the roof finishing with a verge.

Purlins

Purlins are horizontal members introduced to provide an intermediate support to the common rafters, and in turn are supported by gable walls, roof trusses, or by the use of struts supported on partition walls where the rafters are hipped on each end. Care should be taken that the struts, when used, have a good bearing at their feet. Stud partitions, rising from a timber floor below, are not altogether satisfactory.

Spacing of Purlins

The spacing of the purlins is governed to a large extent by the roof covering, and to some extent by the type of roof design. In order that the common rafters may have economical dimensions, the purlins that carry them require to be within a spacing of not more than 10 ft. The purlins should be placed in position before the rafters are fixed, although they are sometimes added afterwards.

Trussed Purlins

Where, owing to the length of the roof, the purlins would become inconveniently large (see Fig. 15), a useful construction that may be used, and might be more often adopted, is the trussed purlin. It will also be found suitable where there is not the necessary room for a principal, and the space inside the roof is to be utilised.

The purlins should extend from wall to wall, and rest on stone templates built into the wall. They should be kept clear of the floor joists, otherwise, when wind pressures come suddenly on one side of the roof,

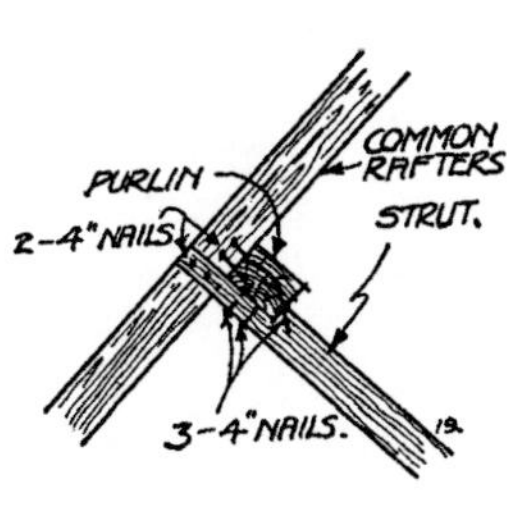

Fig. 13A (*left*).—JOINT BETWEEN RAFTERS, PURLIN, AND STRUT WHERE LATTER IS AT ABOUT RIGHT ANGLES TO RAFTERS

Fig. 13B (*right*).—JOINT WHERE STRUT IS NOT SET AT ABOUT RIGHT ANGLES TO RAFTERS

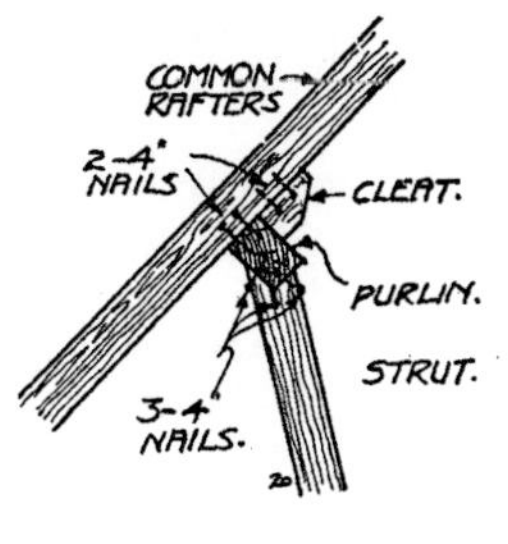

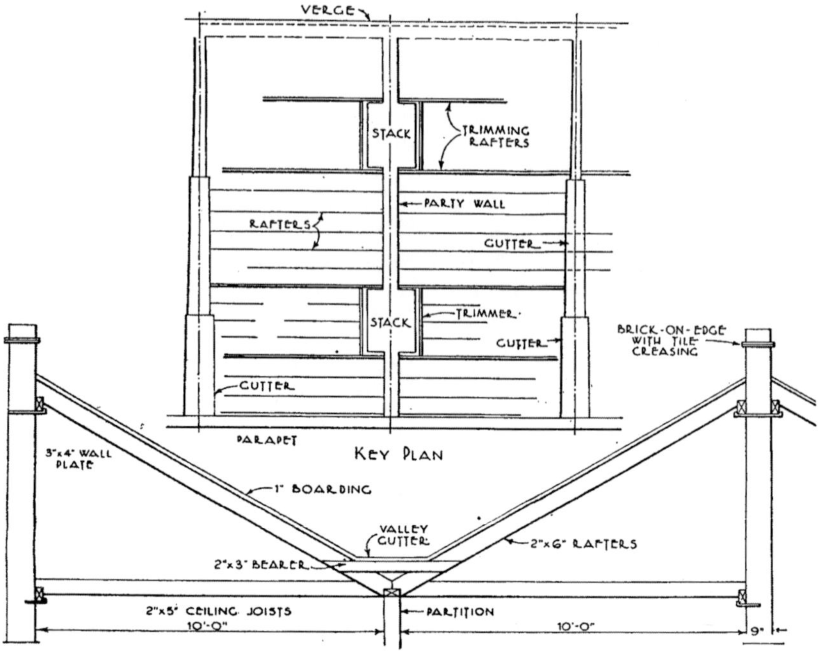

Fig. 14.—VALLEY ROOF

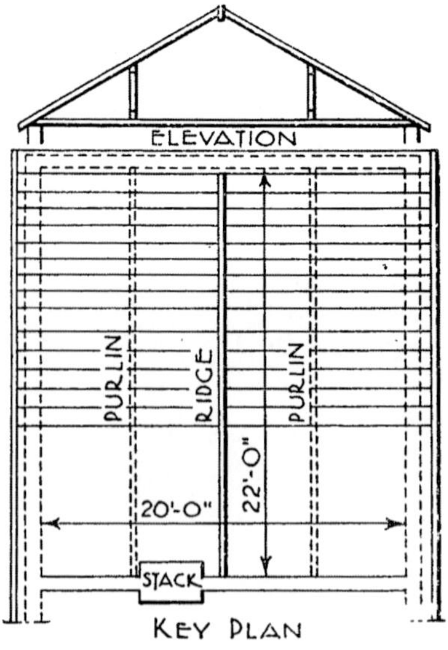

Fig. 15.—PURLIN ROOF

the trussed purlin would consequently be subjected to stresses which might cause sufficient deflection to crack the plaster ceiling below. The bottom member is usually made 4 in. by 4 in., and the struts 3 in. by 4 in., with $\frac{1}{2}$-in., $\frac{5}{8}$-in., or $\frac{3}{4}$-in. bolts, but, if desired, the stresses could be calculated to obtain the exact sizes required (see Fig. 16). The truss usually forms the ashlaring along the sides of the room, and may be filled in with studding to receive the plaster or to be covered with a fibre board.

Roofs Constructed of Short Timbers

Other simple types suitable for comparatively large spans can be built up by placing pieces of various lengths side by side with the use of a truss. In these types it is usual to place the two rafters a few inches apart, to allow for

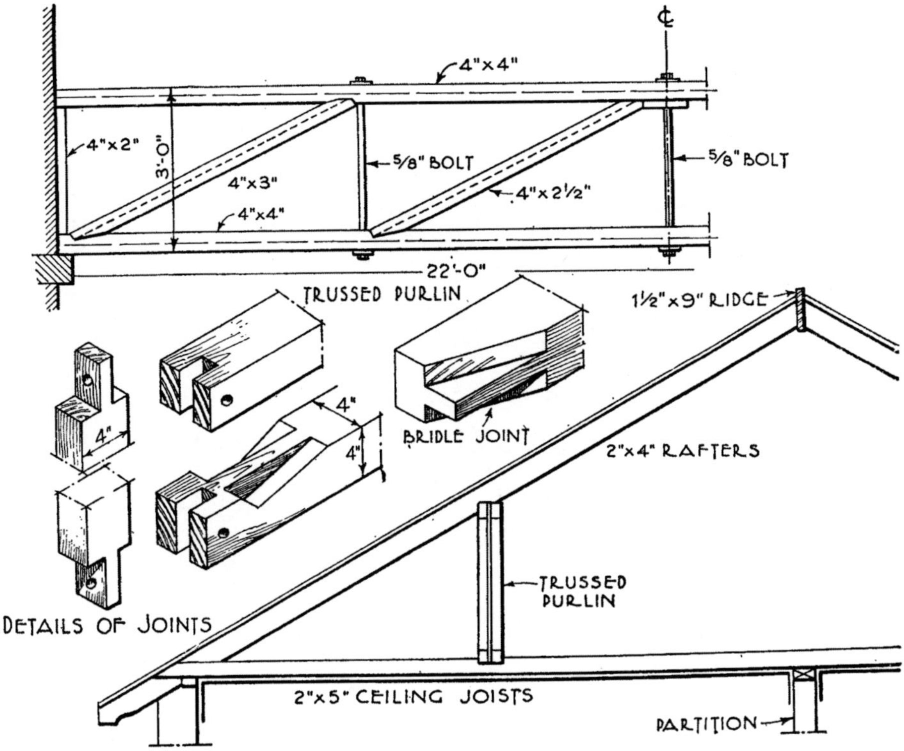

Fig. 16.—Purlin roof

the internal members being fitted securely between the built-up rafters and main tie. The members A, in Fig. 17, are formed at their ends with shoulders which bear on the double rafters and the members, and part formed to extend between the rafter and tie as a tenon. Member B is similar, excepting that at the lower end it notches over the ties and finds a bearing on them. The rafter back near the apex of the truss is formed with two 2-in. by 5-in. pieces placed together, one acting as a joint-plate in lieu of a fish-plate.

When properly fastened together by nailing and bolting, the short pieces are equally as strong as single pieces, and are cheaper and easier to obtain.

Single Mansard Roof

This type is being much used in domestic work. For small spans, it provides an economical means of containing and roofing upper floors and bedrooms. When properly built, it is constructionally sound, and

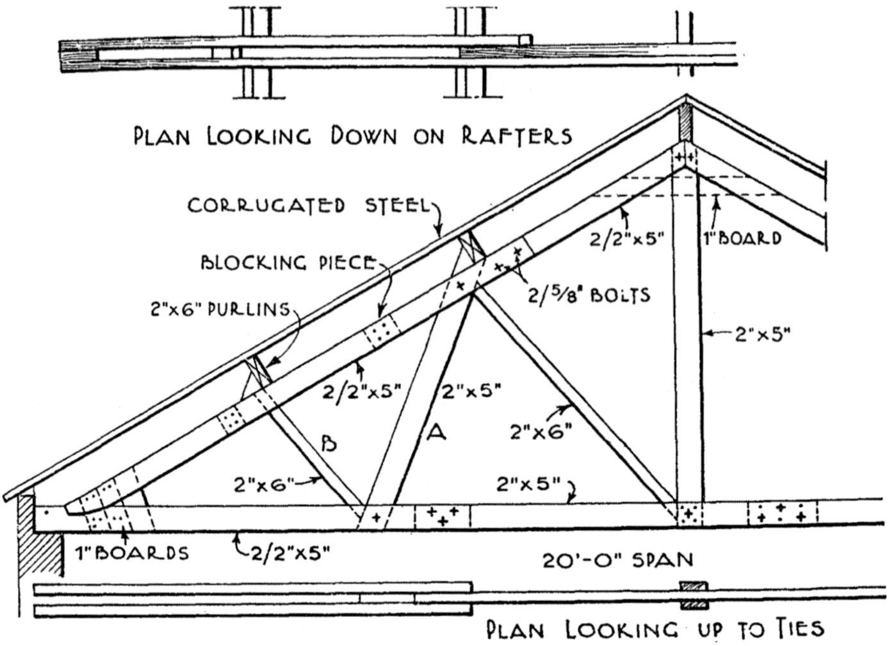

Fig. 17.—Roof construction of pieces of timbers

also capable of pleasing architectural treatment. It is unnecessary to use trusses to support this type of roof over small spans up to 20 ft.

There are many variations in the outline, usually obtained by using one or other of the original French methods of setting out adopted by Mansard, the inventor. Fig. 18 shows the methods used. A semicircle is drawn equal in diameter to the effective span. This is divided into four or five equal parts, giving slightly different outlines. In some cases the lower pitch is inclined at 60° until sufficient headroom is obtained, and then the upper slope inclined at 30°. In modern work the lower slope is usually made 75°, which is the maximum allowable under the Building Acts. For small buildings the lower slope is inclined at 75°, and the upper at from 26° to 45°, according to the covering used.

To obtain the maximum amount of room it is usual to finish with gable-end construction, at least to the height of the lower pitch or kerb. Fig. 19 shows a type commonly adopted, as it gives more floor space. The type shown in Fig. 20 is of stronger construction, but a certain amount of floor space has to be sacrificed. Additional strength can be obtained by nailing cross-pieces between the ashlaring or upright pieces and the

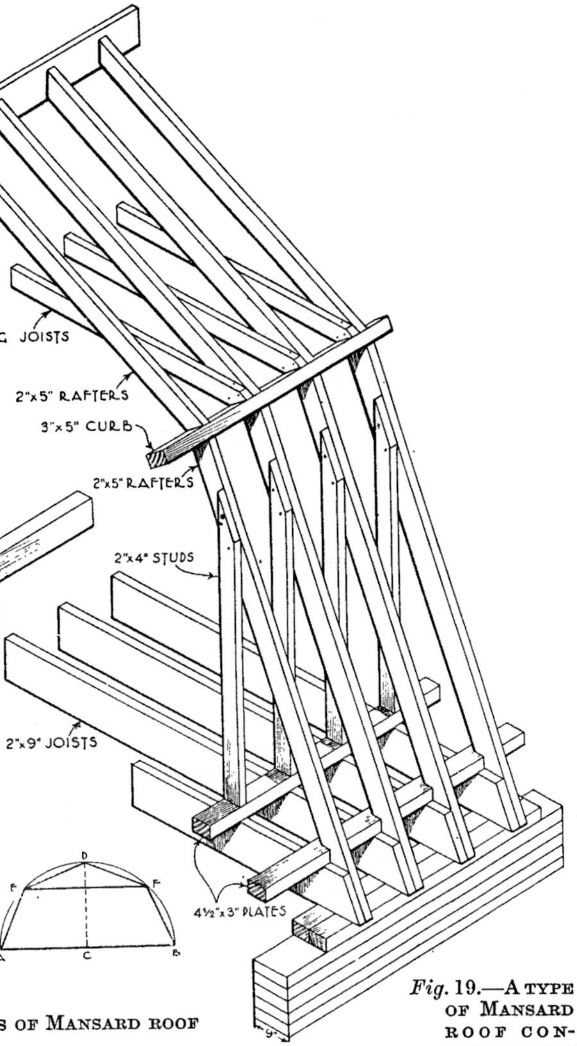

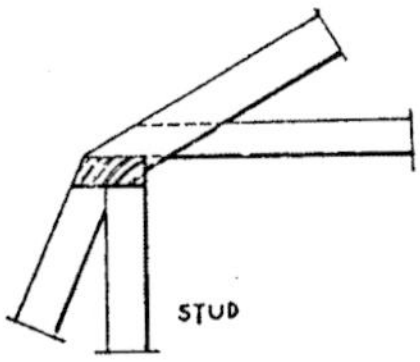

Fig. 20.—ANOTHER TYPE OF MANSARD ROOF CONSTRUCTION

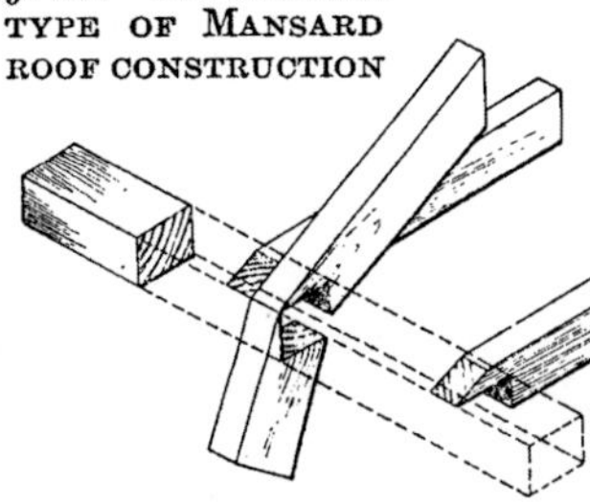

Fig. 21.—ALTERNATIVE METHODS OF FINISHING AT CURBS

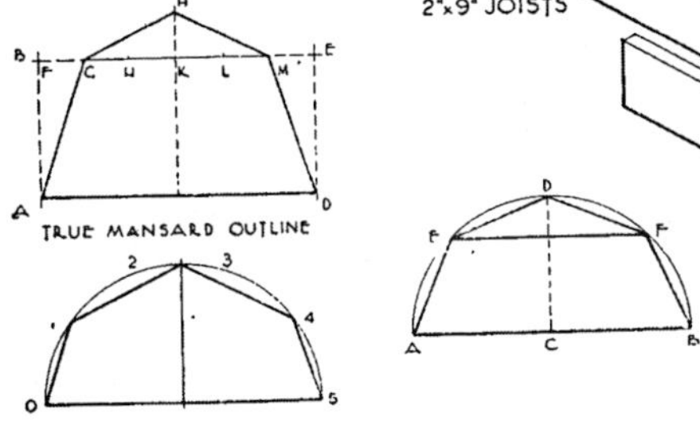

Fig. 18.—TYPICAL OUTLINES OF MANSARD ROOF

Figs. 18 *to* 21.—SINGLE MANSARD ROOF DETAILS

Fig. 19.—A TYPE OF MANSARD ROOF CONSTRUCTION COMMONLY ADOPTED

rafters, usually about every fourth member. This bracing will considerably stiffen the frame and lessen its liability to rack. The rafters should be bird's-mouthed to the wall plate, and also fixed to the side of joists. This will help to stiffen the roof. Every fourth or fifth stud should be tenoned into the cill and curb. Where wind bracing is employed, the

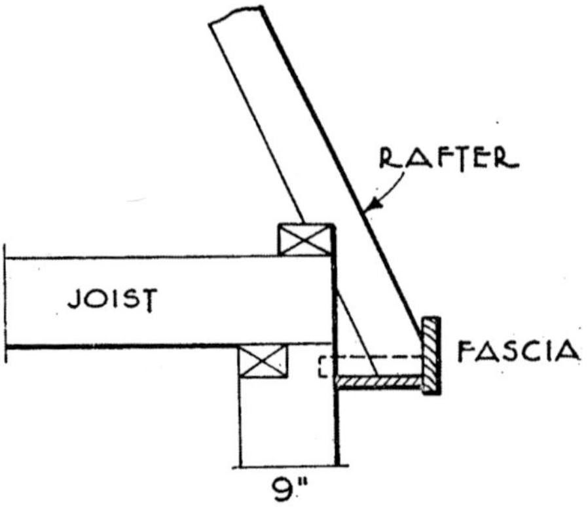

Fig. 22.—EAVES OF SINGLE MANSARD ROOF FINISHED WITH FASCIA

tenoning of the studs may be omitted provided the joints are well made and securely fixed. The floor joists will vary in depth according to span. They are often supported at intermediate points by party walls or partitions.

An alternative method of finish at the curb is shown in Fig. 21, and the finish at the eaves in Fig. 22.

Types of roof construction for larger spans are dealt with in the next chapter.

Eaves Finish for Roofs

Various methods are used for finishing the eaves of roofs. In some cases they are cut off flush with the face of the wall; in others the feet of the rafters project and are wrought and in some cases moulded (see Fig. 23). They may be closed in with soffit boarding fixed to bearers, which gives a neat finish (see Fig. 24). A more elaborate method is shown in Fig. 25, where the soffit has the addition of moulded brackets or consoles.

Another and durable method is to finish the eaves with roofing tiles nailed to the fascia board in the front and resting on a projecting brick course along the wall. Tiles having a continuous nib along the top look best, and therefore make a more suitable finish than the un-nibbed variety. Fig. 26 illustrates this treatment.

Tilt at the Eaves

Tiled roofs do not require any exaggerated tilt at the eaves such as is given by the use of sprocket

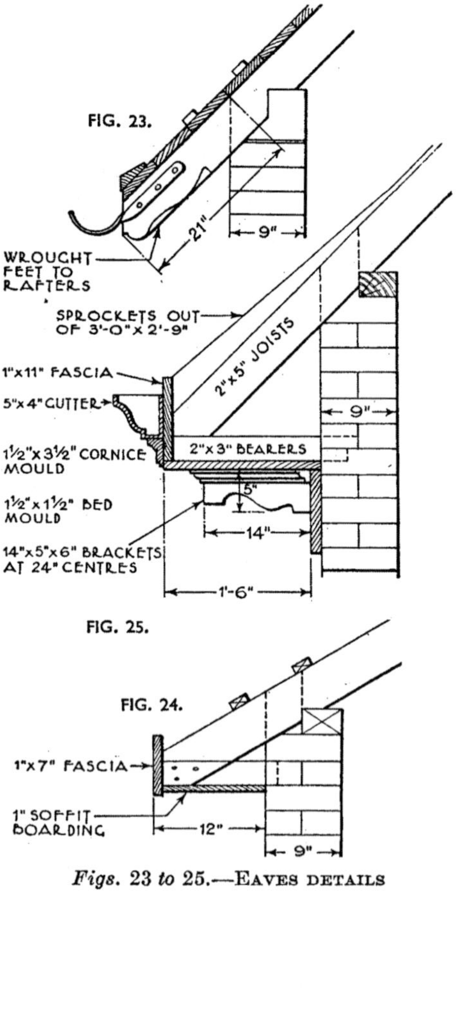

Figs. 23 *to* 25.—EAVES DETAILS

pieces (see Fig. 25), but where such are employed care should be taken that the difference in pitch between the sprocket and the roof proper is not too great, nor that the pitch adopted for the covering at this point is too flat where a large volume of water has to be dealt with.

Sprockets may be fixed to the top of the rafters or nailed to their sides.

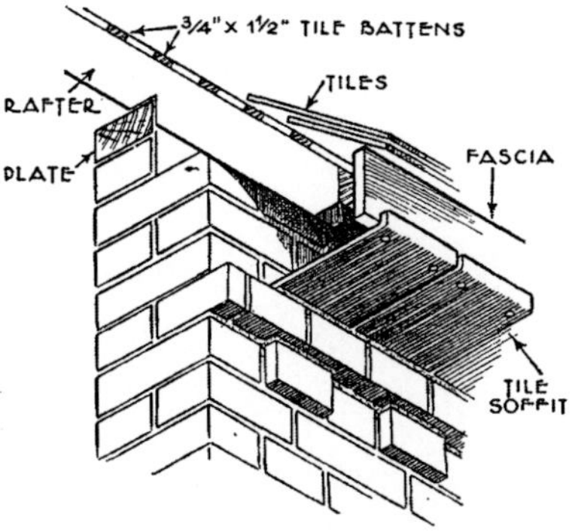

Fig. 26.—EAVES FORMED WITH FASCIA BOARD AND TILE SOFFIT

Precast Concrete Roofing Units

A recent development in the construction of roofs of comparatively small span is the introduction of precast concrete units suitable for supporting lightweight roof coverings. This type of construction is likely to develop rapidly to meet the urgent need for speed in erection and saving in timber occasioned by the post-war construction programme.

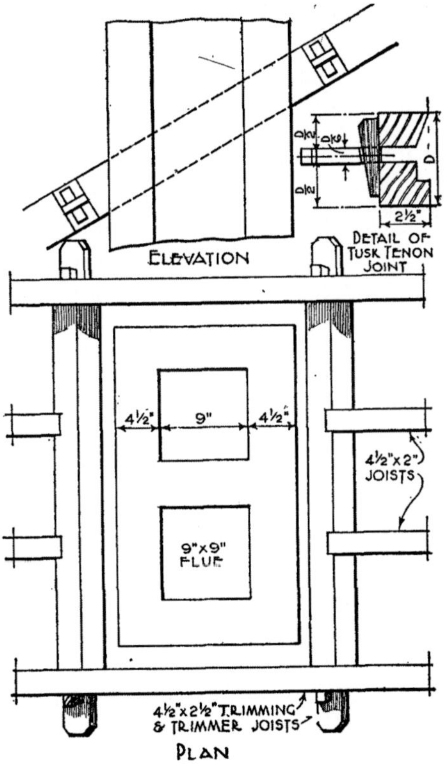

Fig. 27.—TRIMMING TO CHIMNEY STACK

TRIMMING

Chimneys, etc.

Where chimneys, dormer windows, skylights, or other openings occur in a roof, it is generally necessary to trim the rafters in order to leave sufficient open space. The trimmers are tenoned into the trimming rafters, and these are generally made ½ in. thicker than the ordinary rafters. The trimmed rafters are notched to the trimmers. A frame is thus provided in the roof surface without any rafters crossing it (see Fig. 27).

Dormer Windows

The trimming for a dormer window is shown in Fig. 29.

Owing to building laws the projection of dormers is in some

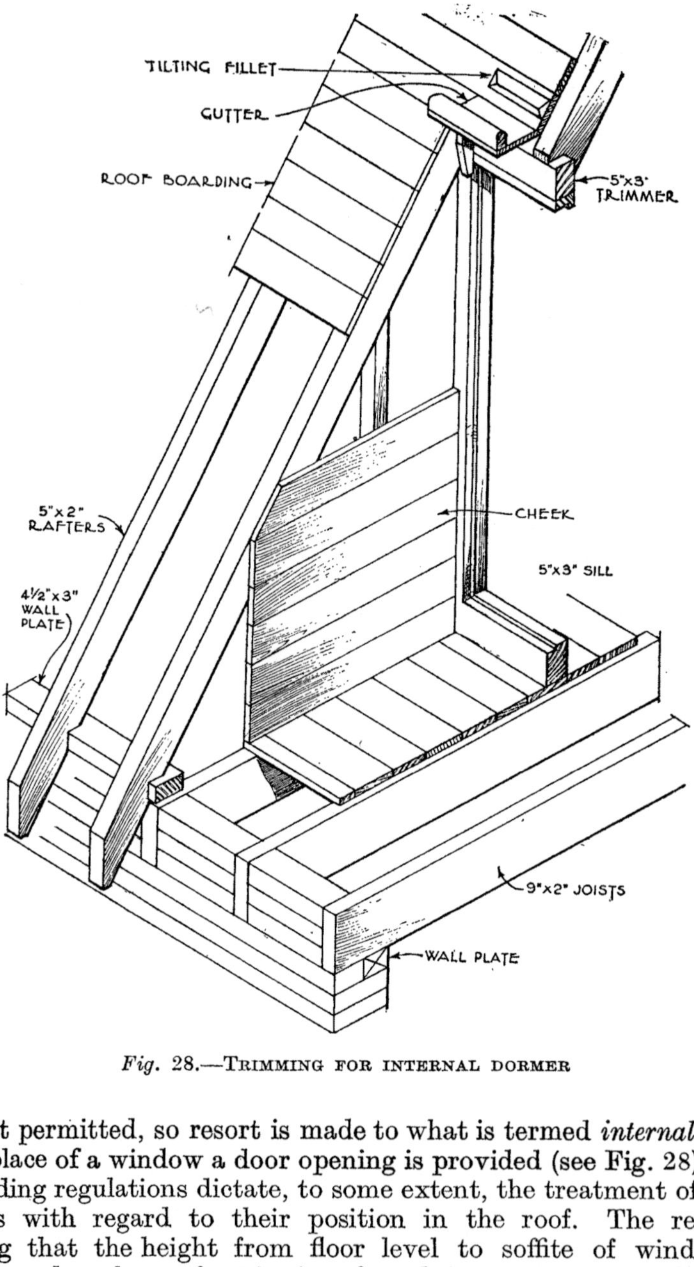

Fig. 28.—Trimming for internal dormer

cases not permitted, so resort is made to what is termed *internal dormer*, and in place of a window a door opening is provided (see Fig. 28).

Building regulations dictate, to some extent, the treatment of dormer windows with regard to their position in the roof. The regulation requiring that the height from floor level to soffite of window-head shall be not less than 6 ft. 6 in. is enforced, in most areas, in the case of dormer windows, as is also the regulation that the ashlaring in rooms

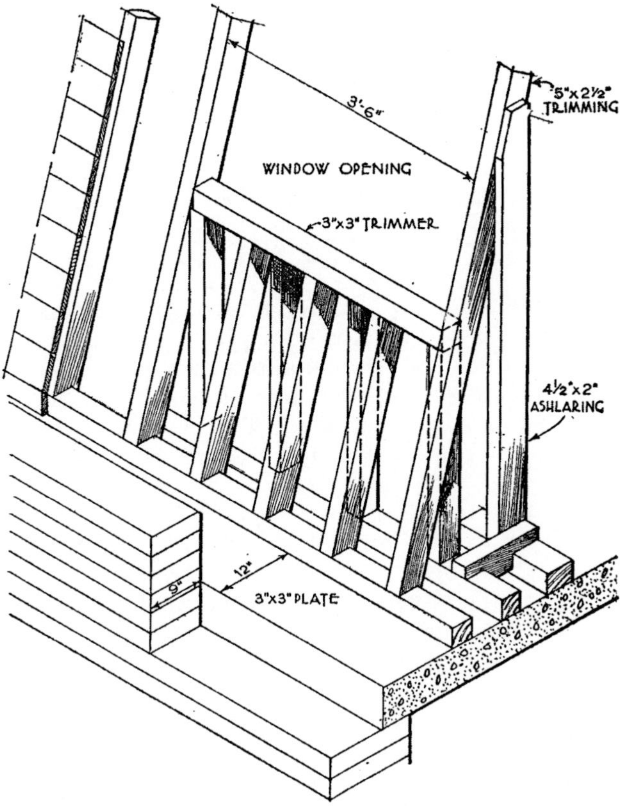

Fig. 29.—TRIMMING FOR DORMER WINDOW

occupying part of the roof space shall have a minimum vertical height of 4 ft. 6 in.

The combination of these two requirements results in a reasonably attractive appearance both externally and internally.

From the æsthetic standpoint dormer windows look best if they project above the roof slope either one-third or two-thirds their own height. They do not look so well if the projection approximates to half their own height.

Chapter II

TRUSSED ROOFS

FOR spans of more than 20 ft. a truss becomes necessary, and what is known as a *king-post truss* is employed. It consists of a tie-beam, two principal rafters, two struts, and a king-post. As a rough guide we may take it that the tie-beam requires a support every 15 ft., and the principal rafter a support every 8 ft. According to this rule this type of truss is suitable for spans up to 30 ft. (see Fig. 1).

Principle of King-post Truss

The king-post supports the tie-beam at the centre, and although a vertical member it is subjected to tensional stress alone. The principal rafters are in compression, and are supported in the middle by the struts from the base of the king-post. These also are in compression. The tie-beam is in tension, but where it carries a ceiling load it will also be subjected to transverse bending.

The stress in the principal rafters tends to bend, and consequently, as with the tie-beam, their depth must be greater than their width or thickness. The struts, having no definite transverse strain, should be made as nearly as possible square in section. The purlins are supported by the truss, and in turn support the roofing, i.e. common rafters, boarding, battens, slates, or tiles.

The dimensions given in the Table below are minimum sizes, and may be increased if desired. This has been done on the diagrams, Fig. 1.

DIMENSIONS OF TIMBERS FOR KING-POST ROOF TRUSSES
10 FT. APART, PITCH 30°, SLATED

Span	*Thickness of All Truss Members*	*Minimum Breadth on Elevation*			
		Tie-beam	*Principal Rafter*	*King-post*	*Struts*
ft.	*in.*	*in.*	*in.*	*in.*	*in.*
20	4	8	4	3	2
22	4½	9	4	3½	2¼
24	5	10	4¼	3½	2½
26	5½	10	4¼	4	2¾
28	6	11	4½	4	3
30	6½	12	4½	4½	3¼

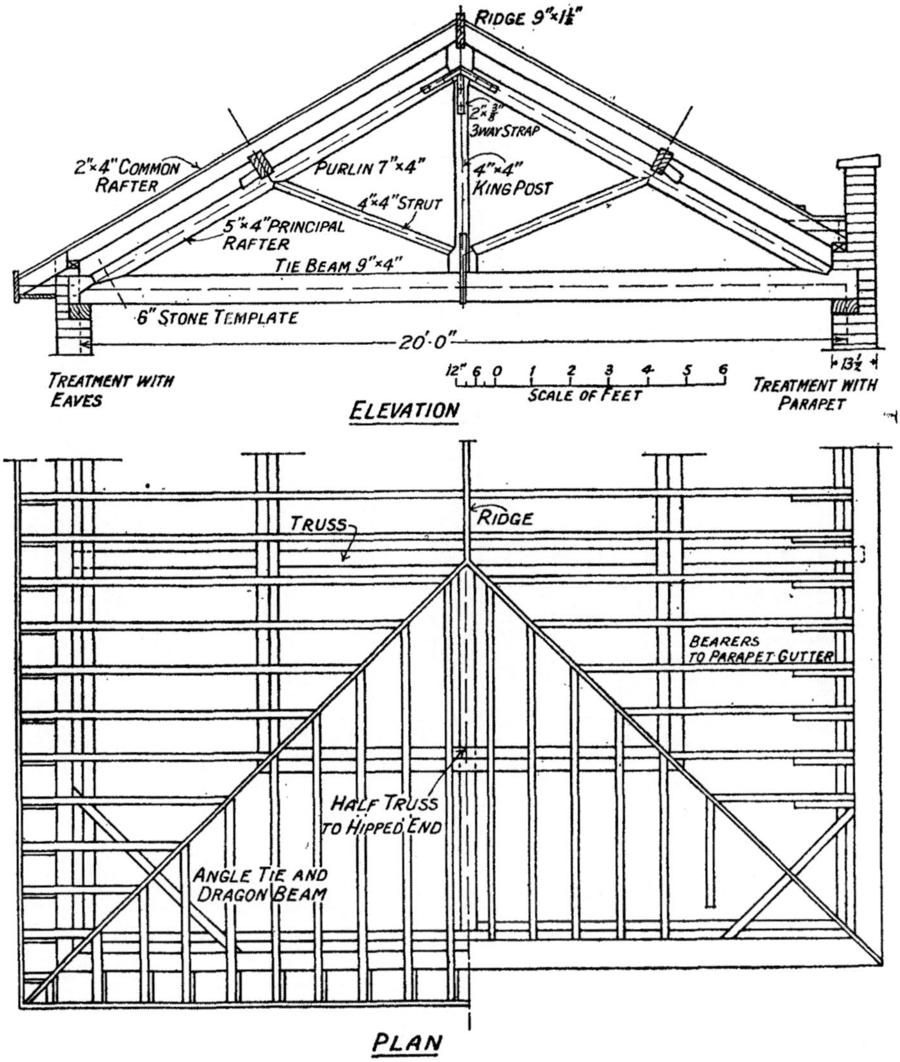

Fig. 1.—King-post roof truss with hipped end

All the members of a king-post roof truss are made the same thickness with the exception sometimes of the struts. This allows the iron straps to be bolted to the surface to hold the joints together.

The Purlins

The purlins are generally placed over the top of the struts, and as there are only two struts in a king-post truss the purlins are consequently

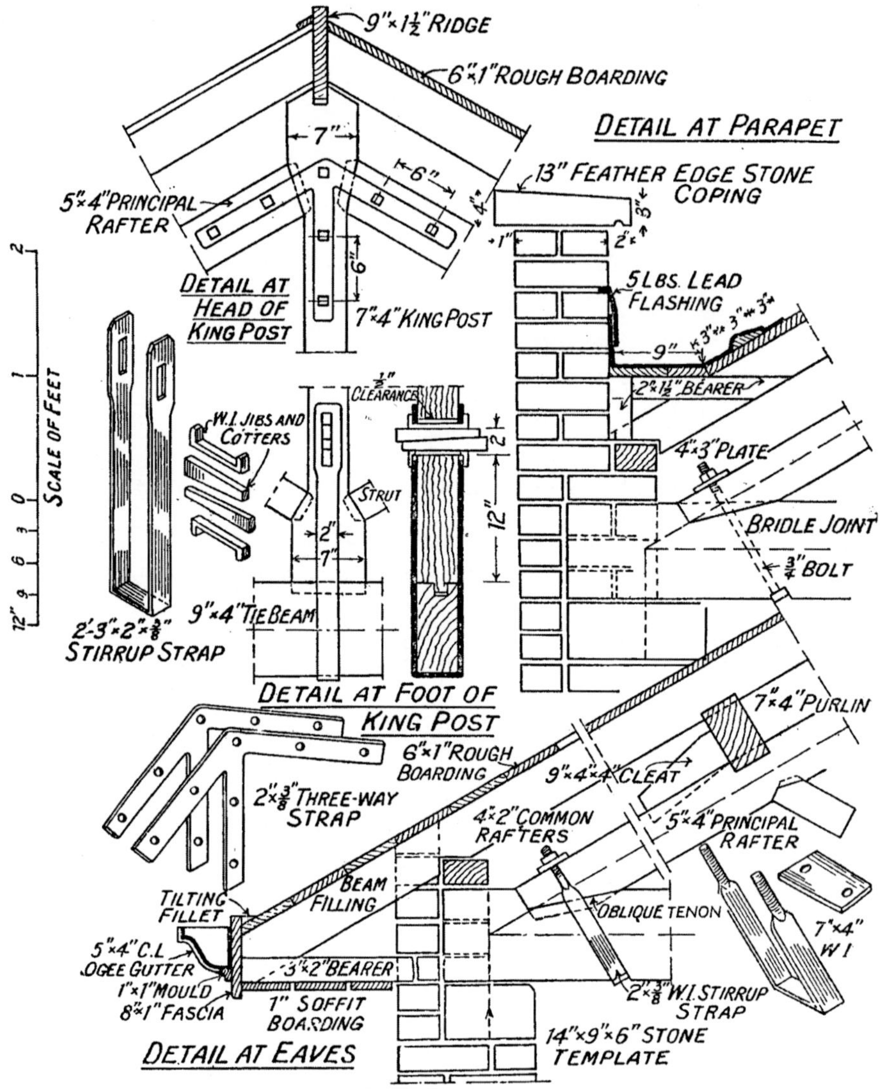

Fig. 2.—KING-POST DETAILS

Showing details at head and foot of king-post, eaves, and parapet.

two in number. The purlins are cogged or recessed about ½ in. or so, to fit over the principal rafters, and are fixed by nailing. They are further secured and supported by cleats to prevent them from turning over or slipping down. Details of joints are shown in Fig. 2.

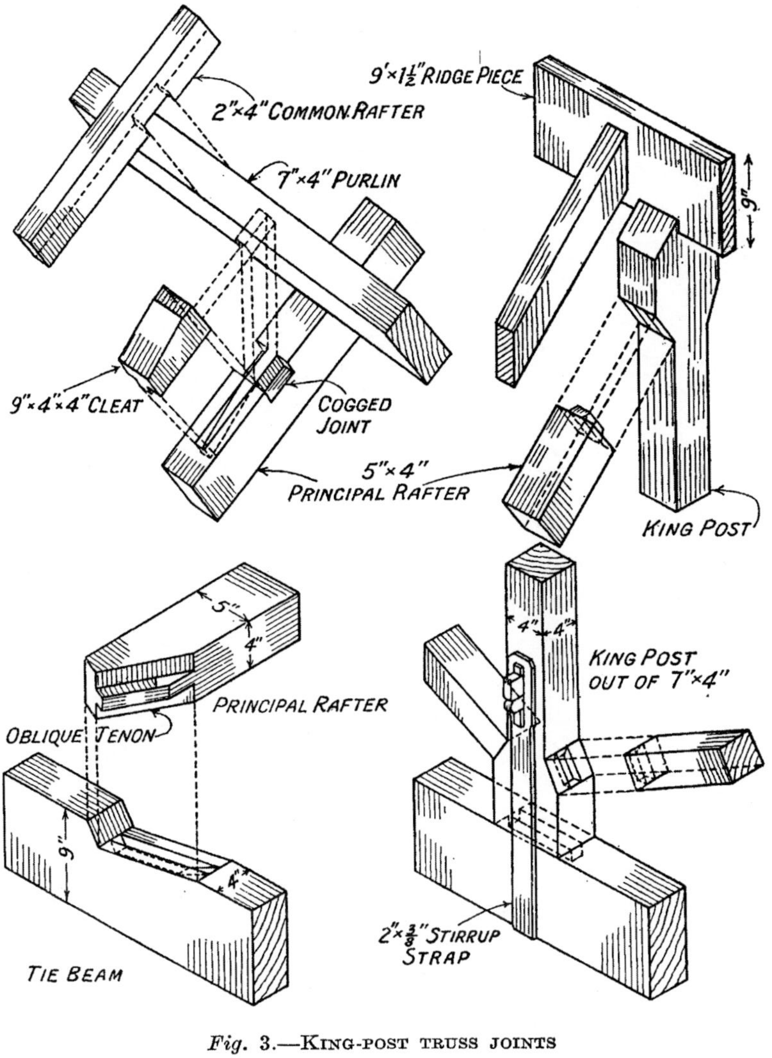

Fig. 3.—King-post truss joints

To form an abutment as nearly as possible at right angles to the inclination of the member, the king-post is reduced in size to about half its width in the middle, and to secure the joints further, iron straps are used (see Fig. 2).

Joints

The joint between the top of the principal rafter and the king-post is strengthened and made more secure with two three-way straps and

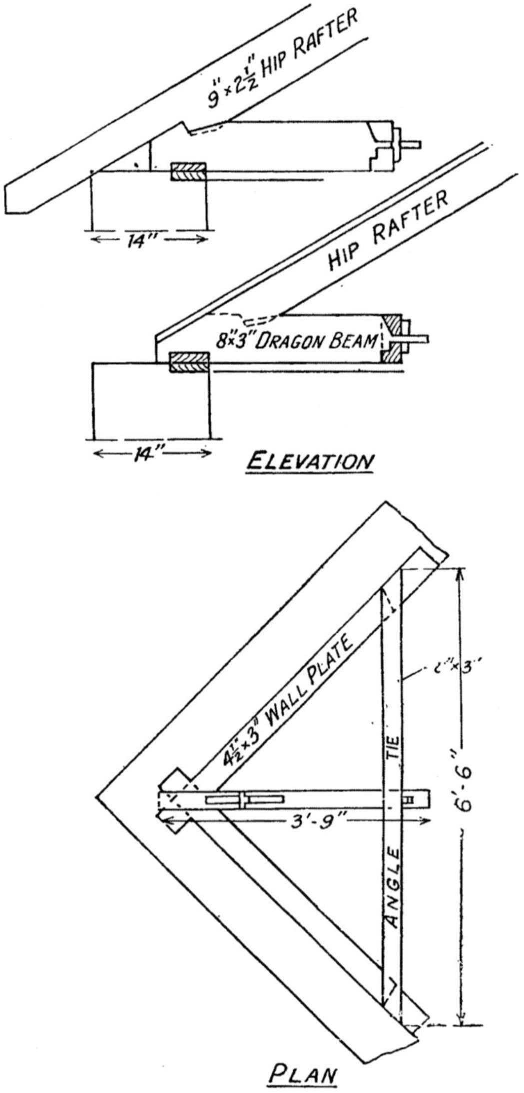

Fig. 4.—ANGLE TIE AND DRAGON BEAM

½-in. bolts, as shown. The joint between the king-post and the tie-beam is connected by a stirrup and tightened by folding wedges, sufficient clearance being allowed in both the stirrup and the post for this purpose (see Fig. 2). Heel-straps or bolts are used to hold the principal rafters to the tie-beam, and the principal rafter is formed by an oblique tenon or a bridle joint. Isometric sketches of the joints are shown in Fig. 3.

The trusses should be put together with the king-post slightly short, and the struts slightly full, to length. In small roofs what are known as *dragon beams* and angle ties are employed to prevent the outward thrust of the hip rafter from acting directly on the corner of the building (see Fig. 4).

Sizes of Various Members

To find the sizes of the various members the following rules may be employed. For the thickness of all members, take ¼ span in feet — 1, = thickness in inches. The breadth of the members on the elevation will be : tie-beam, ½ span in feet — 2 ; centre of king-post, ⅙ span in feet. Make king-post ends twice width of centre less ½ in. ; struts 2 in. for 20 ft., increasing ¼ in. in width for every 2 ft. Reduce thickness of truss by ½ in. and depth of tie-beam by 1 in. if there is no ceiling.

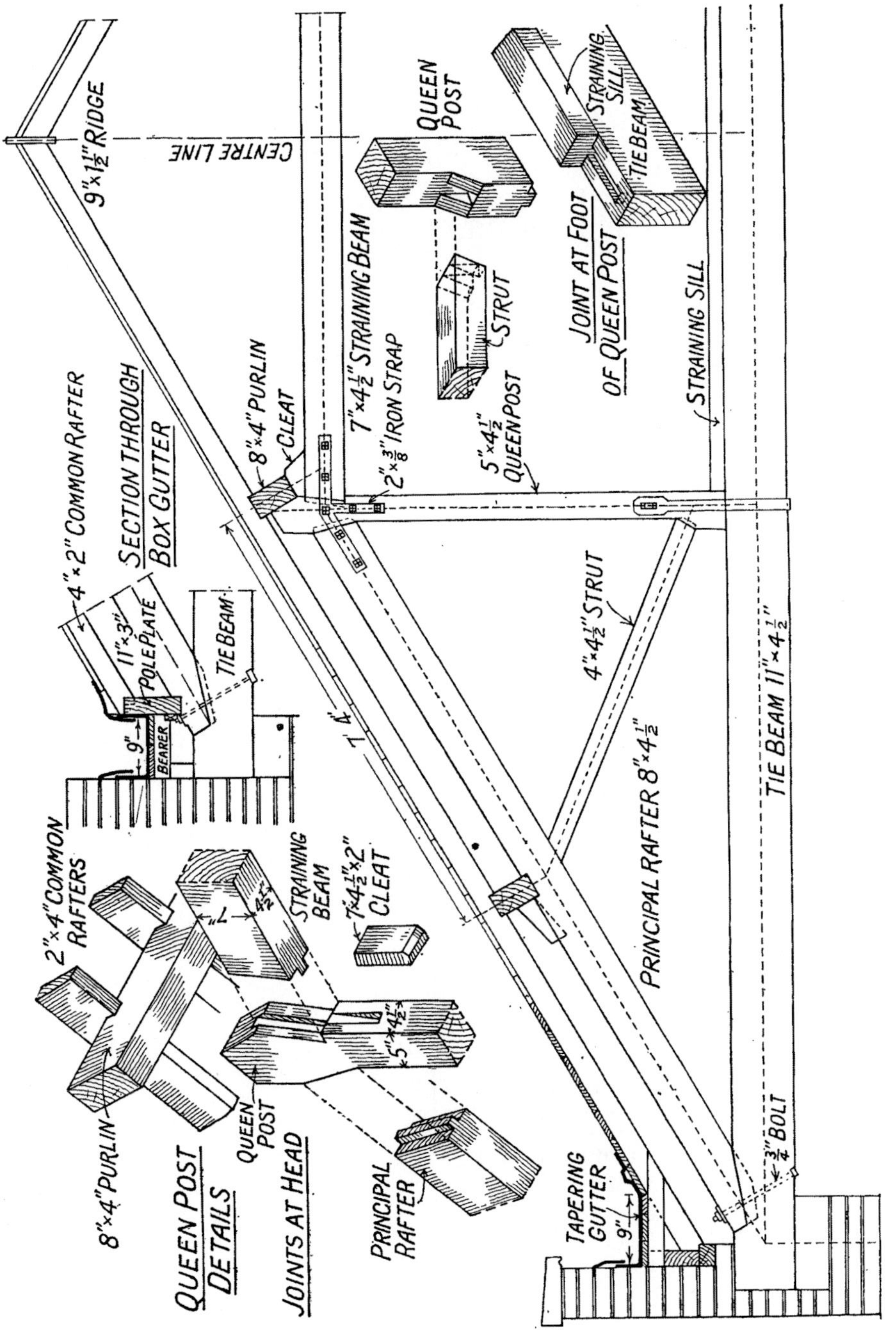

Fig. 5.—Queen-post Roof Russ

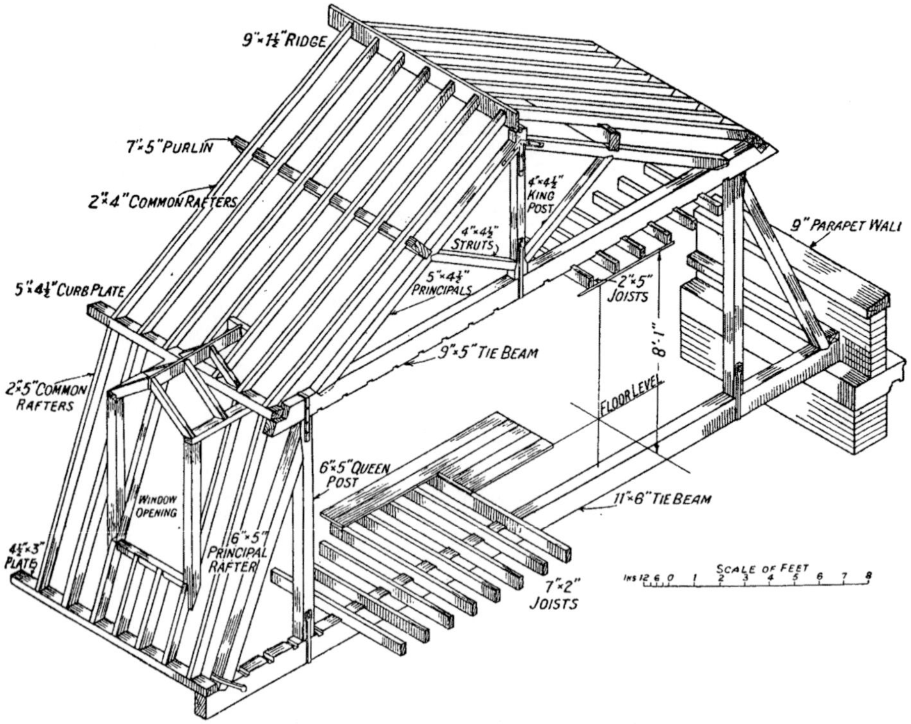

Fig. 6.—Isometric sketch of Mansard roof, showing use of combined king- and queen-post trusses as supports

DIMENSIONS OF TIMBERS FOR QUEEN-POST TRUSSES 10 FT. APART, PITCH 30°, SLATED

Span	*Thickness of Truss Members*	*Breadth on Elevation*					
		Tie-beam	*Principal Rafter*	*Queen-post*	*Struts*	*Straining Beam*	*Straining Cill*
ft.	*in.*	*in.*	*in.*	*in.*	*in.*	*in.*	*in.*
32	4	9	4½	4	2	6	3
34	4½	10	5	4	2	6½	3½
36	5	10½	5½	4½	2	7	4
38	5½	11	5½	4½	2¼	7½	4
40	5½	11½	6	5	2½	8	4½
42	6	12	6	5	2½	8	5
44	6	12	6½	5½	2½	8½	5

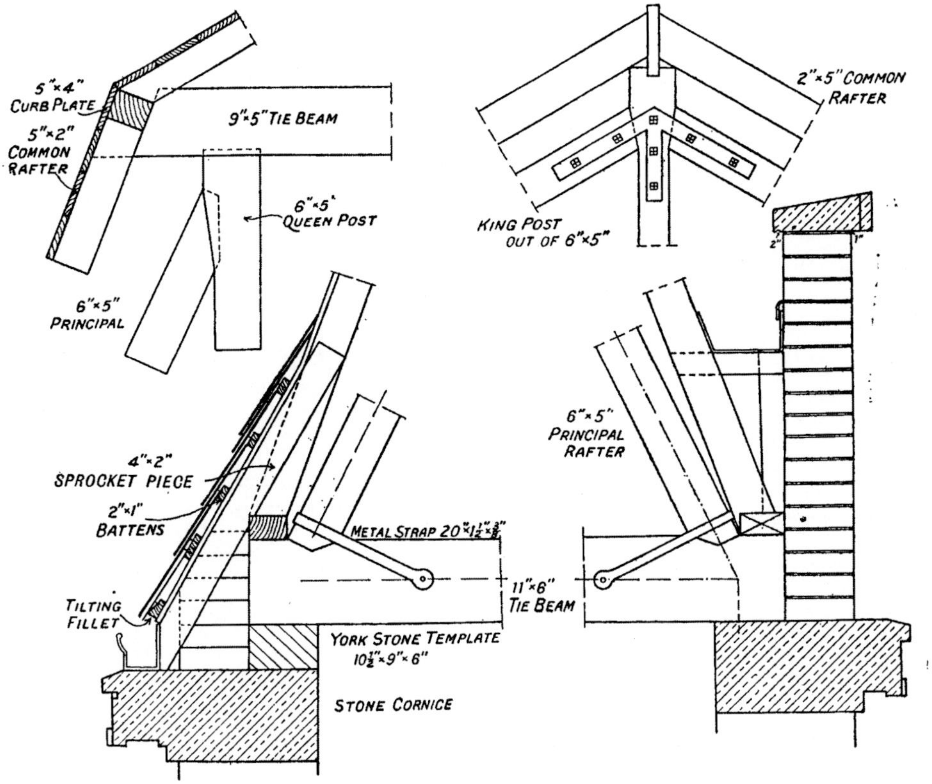

Fig. 7.—MANSARD ROOF DETAILS

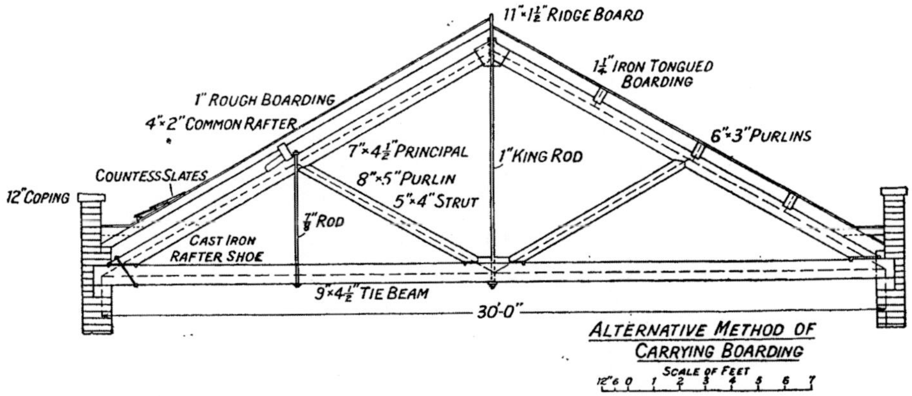

Fig. 8.—COMPOSITE TRUSS
Showing alternative methods of carrying boarding.

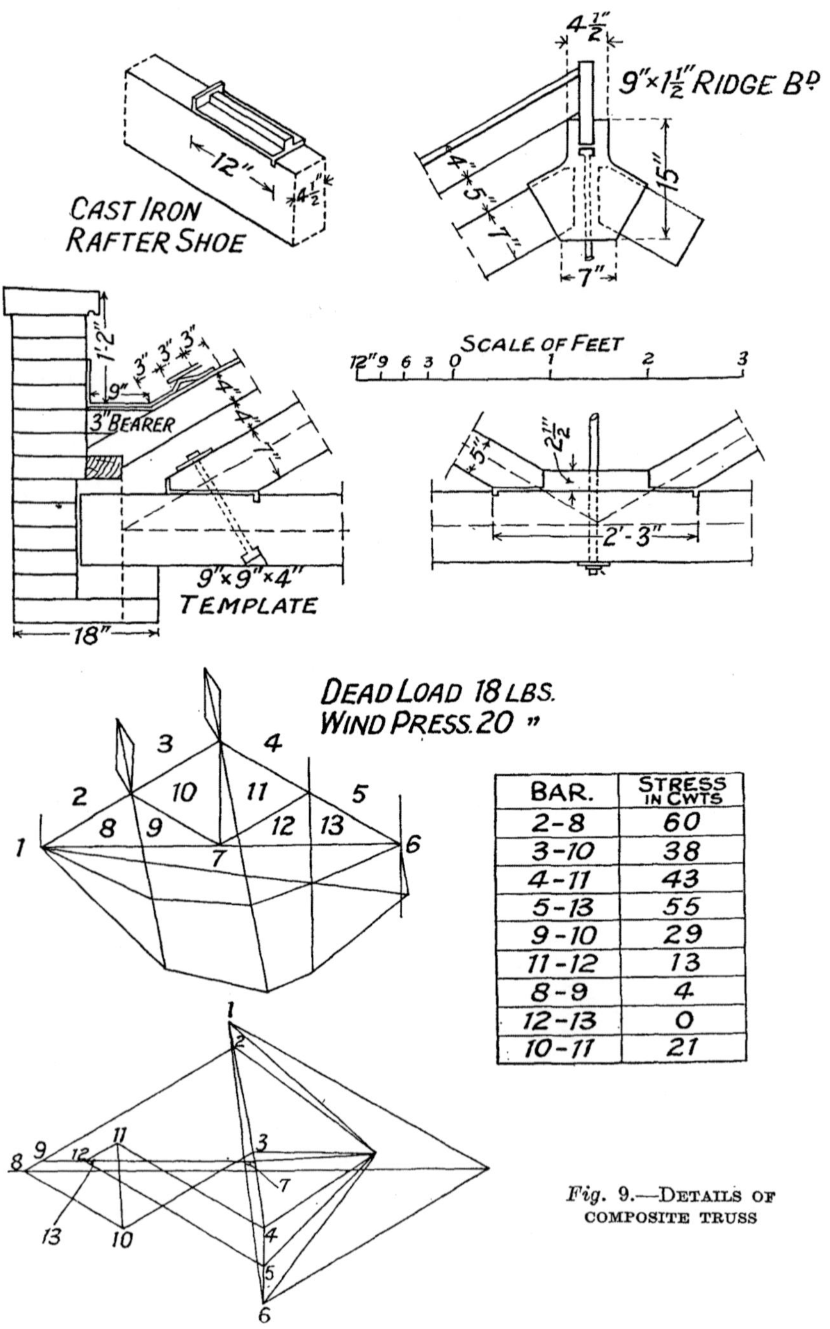

BAR.	STRESS IN CWTS
2-8	60
3-10	38
4-11	43
5-13	55
9-10	29
11-12	13
8-9	4
12-13	0
10-11	21

Fig. 9.—DETAILS OF COMPOSITE TRUSS

Queen-post Trusses

These are used for spans of from 30 ft. to 45 ft., the tie-beam and rafters being supported at two intermediate points in their length, instead of only at one. The two queen-posts are connected at their tops by a horizontal member called a *straining beam*, and the junction of these affords a support at one point for the rafters (see Fig. 5). The principal rafter ends at this point, and is not carried to the ridge as in the king-post roof truss. A support for the middle of the principal rafter is obtained by the strut which springs from the base of the queen-post. The purlin over this point occurs at about a third of the distance along the common rafter, the second purlin at the top of the queen-post being another third, which provides the necessary support to the common rafters.

The thrusts from the struts are resisted by a tenon on the foot of each post, and a member called a *straining sill*. Many of the details of this truss are the same as those of the king-post truss previously described and illustrated (see Fig. 5).

Queen-post roof trusses are not much in use at the present time, owing to the increased use of steel for large-span roofs, but they can be used with advantage where a lantern light has to be fixed in the centre. The Table on page 24 gives the scantlings for queen-post roof trusses.

Mansard or Curb Roof

The Mansard or curb roof is essentially one with two pitches, and is usually employed as a means of economising space. In roofs of ordinary pitch a great amount of the space is lost, due to the sharp pitch of the sides, but in the Mansard roof this is overcome by making the sides of the lower pitch nearly vertical. The upper slope is then made relatively flat to avoid an acute ridge. Fig. 6 is an isometric view of this roof, and Fig. 7 shows the details of the joints.

Composite Roof Trusses

Roofs formed or built up of a combination of timber and steel members are known as composite trusses, the compression members being of timber and the tension members of steel, which from its great tensile strength is more suitable, but these roofs are not very frequently used (see Fig. 8). Alternative methods of construction are illustrated.

Chapter III

FLAT ROOFS

IN practice, flat roofs are formed with timber, concrete and steel, reinforced concrete, or hollow tiles as a base, and covered with asphalte lead, copper, or zinc, or concrete slabs over asphalte, or asbestos-cement tiles on asphalte. The construction is similar to floor construction, but a slight fall is necessary to drain off the water.

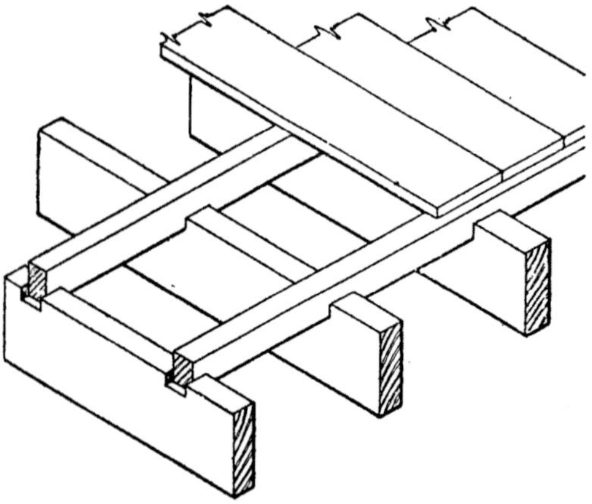

Fig. 1.—FLAT ROOF—JOISTS PARALLEL TO FLOW

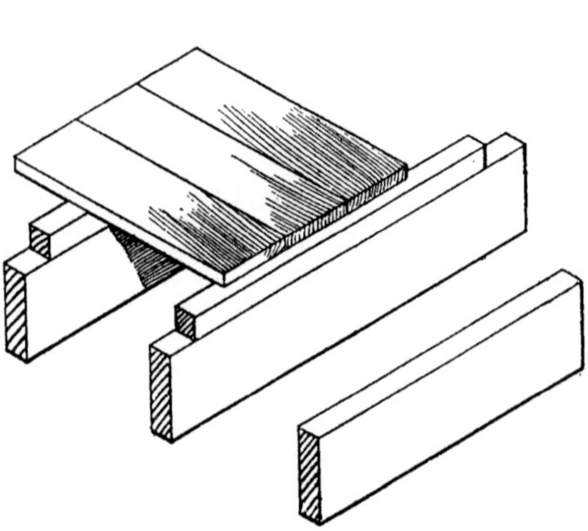

Fig. 2.—FLAT ROOF—JOISTS AT RIGHT ANGLES TO FLOW

Timber Construction

When timber is employed the joists are furred to falls (Fig. 1). Where the span of the joists exceeds about 16 ft., rolled-steel joist binders are introduced to form an intermediate support for the joists. For lead-covered flats, the roof is given a fall of $1\frac{1}{2}$ in. in 10 ft., or 4°. The boards should run in the direction of the flow, because if placed at right angles to it, and the boards should warp slightly, hollows would form in which the water may lie (Figs. 1 and 2). For good-class work the boarding should be in narrow widths, and $1\frac{1}{4}$-in. nominal thickness, and the roof covered with felt to give a smooth surface for the lead or other covering.

Lead Flats—Drips

The joint along the end of the lead sheet is formed by

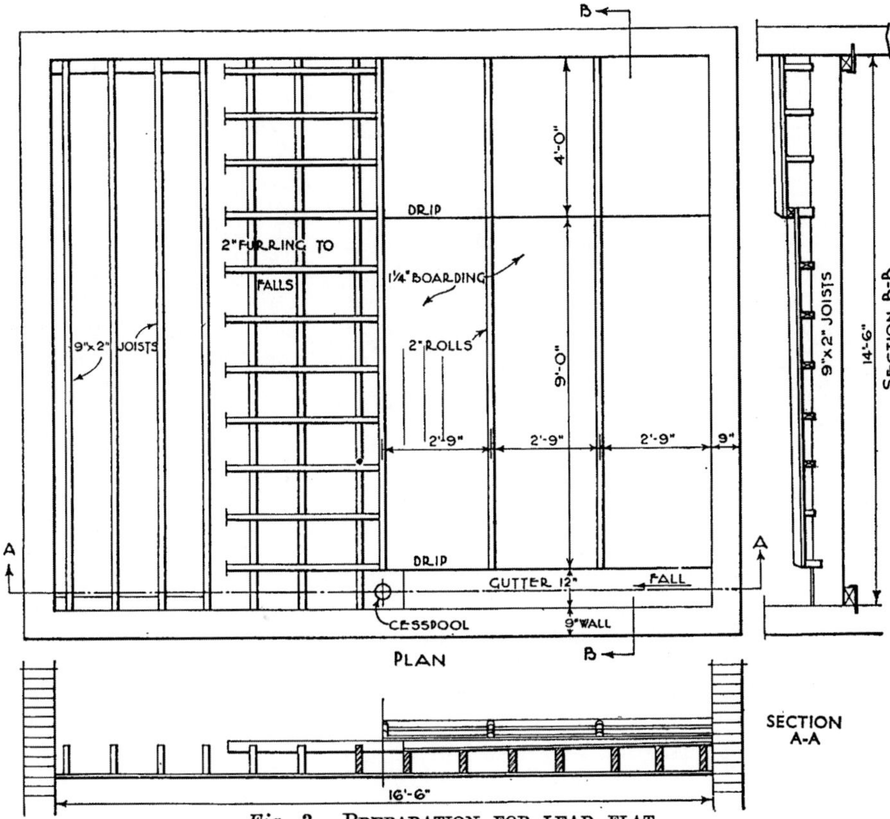

Fig. 3.—Preparation for lead flat

means of a *drip*—an abrupt change in the level of the roof, usually 2½ in. deep. The face of the drip may be vertical (see Fig. 4), or an angle fillet may be introduced.

The end of the boarding at the top is rebated to a depth equal to the thickness of the sheet, and a width of 1 in. or 1½ in. The upper end of the lower sheet is dressed over the drip and turned down into the rebate, and fixed with copper nails. The upper sheet is next dressed down over the drip, and is either finished about ½ in. short of the lower flat or carried down and dressed about 1 in. on the flat below. The amount of lead required to form a 2½-in. drip is 7 in., and this amount, deducted from the maximum length of sheet it is desirable to use, i.e. 10 ft., gives the distances between the drips or length of bay at 9 ft. 5 in.

Rolls for Lead

The joint along the edge of the sheet and parallel to the fall is formed by means of a roll, over which the lead sheets are dressed (see Fig. 5). The edge of one sheet, or *undercloak* as it is called, is dressed about two-thirds round the roll, and copper nailed. The adjoining sheet, the *overcloak*, is then dressed over the roll, covering the edge of the first sheet and finishing 1 in. on the flat. Allowing 3 in. for the undercloak and 7 in. for the overcloak, the maximum distance from centre to centre of the rolls is 2 ft. 10 in., which allows two sheets being cut in the width of the roll.

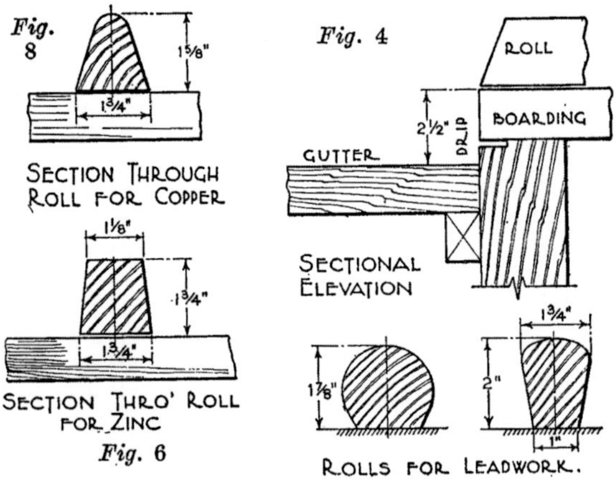

Figs. 4, 5, 6, *and* 8.—DETAILS OF ROLLS FOR SHEET METAL

Zinc Roofing

For zinc work the common sizes of sheets manufactured are 7 ft. by 3 ft. and 8 ft. by 3 ft.

Allowing 6 in. for the drip and 3 in. for the roll, the size of the bays will be 7 ft. 6 in. by 2 ft. 9 in.

The rolls are trapezoidal in form, splayed on their sides from $1\frac{5}{8}$ to $1\frac{3}{4}$ in. at the bottom to a width of $1\frac{1}{8}$ in. at the top, the height being the same as width at bottom (see Fig. 6).

Copper Roofing

For copper roofing the sheets employed measure 5 ft. 3 in. by 2 ft. 8 in., and 4 ft. by 3 ft. 6 in., both having the same area. After making allowances for roll and drip, the bays will measure 4 ft. $10\frac{1}{2}$ in. by 2 ft. $4\frac{1}{2}$ in. when using a 5-ft. 3-in. by 2-ft. 8-in. sheet (Fig. 7). The rolls are made conical shape in section, measuring $1\frac{3}{4}$ in. at the bottom and $1\frac{5}{8}$ in. or $1\frac{3}{4}$ in. in height (see Fig. 8).

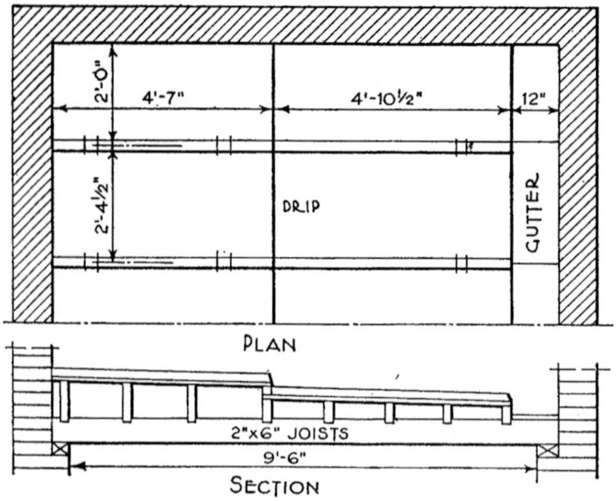

Fig. 7.—PREPARATION OF ROOF FOR COPPER COVERING

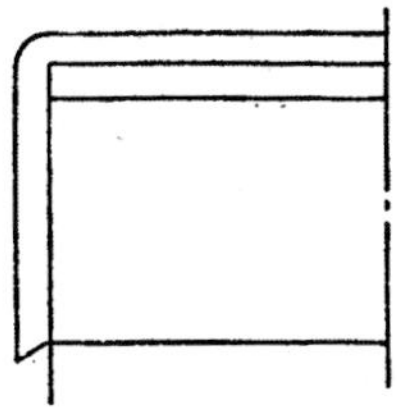

Fig. 9.—ASPHALTE APRON WITH UNDERCUT DRIP

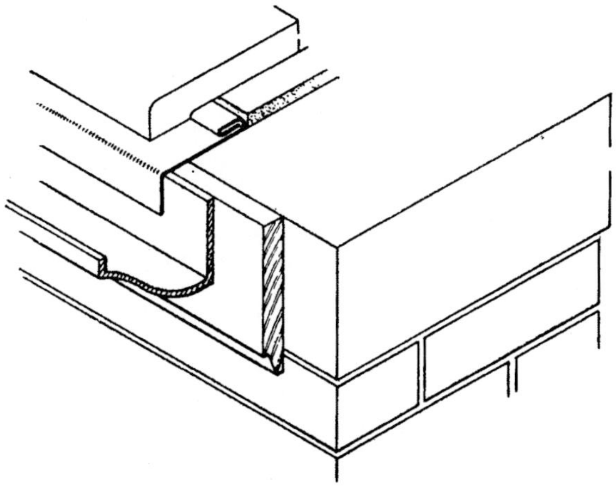

Fig. 10.—ASPHALTE DRESSING TO WELTED LEAD OR COPPER APRON ON REBATED EDGE

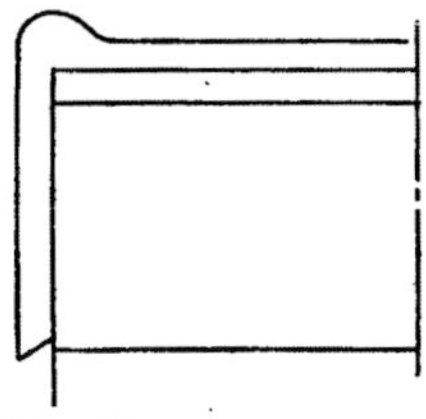

Fig. 11.—ASPHALTE FINISH AT VERGE

Asphalte Roofing

This material can be laid on concrete or boarding ; when laid on boarding an underlay of felt must be employed, but the use of wire or expanded metal lathing in addition is not considered essential. For flat roofs the surface should be finished to a fall of 1 in 10. On sloping roofs the surface should be formed with grooves to provide a key, as the material is apt to run if exposed to the sun. The grooves are formed by fixing 1½-in. by ⅜-in. battens horizontally along the roof at 18-in. centres, and then rendering the surface level with the top of the battens in cement and sand (1 to 2). The battens are removed after the cement has set, leaving grooves in the surface. Where a flat roof finishes with an eaves gutter the asphalte is dressed over on to the face of the wall to form an apron (see Fig. 9), the lower edge being splayed back to form a drip. An alternative method is to dress the asphalte to a welted lead or copper flashing on a rebated edge of the flat, the asphalte being set back ¾ in. from the edge (see Fig. 10). Where verges occur on a concrete pitched roof, the edge is slightly tilted to divert the water from the edge (see Fig. 11).

Junction of Roof with Lantern Lights

Where a flat roof abuts against a lantern light, a curb is formed in concrete or wood, the asphalte carried up the face of the curb and finished

Fig. 12

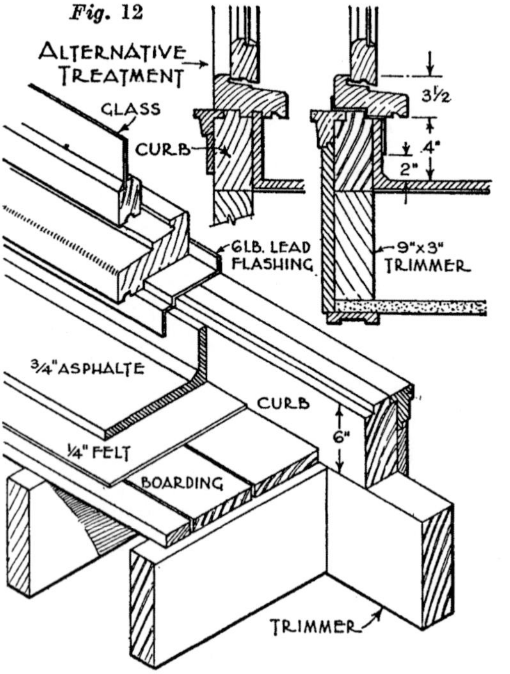

Fig. 13.—Treatment of Timber Roof Abutting Lantern Light

in a groove in the under side of the cill (see Fig. 12), or it may finish against the under side of the cill with the addition of a lead flashing, a groove being formed in the cill for the purpose (Fig. 13). For a concrete curb the asphalte is taken up the face of the curb under the cill, and finished against the water bar (see Fig. 14).

Steel and concrete and glass are now largely used in the construction of lantern lights.

Fig. 14 (*below*).—Treatment for Concrete Roof Abutting Lantern Light

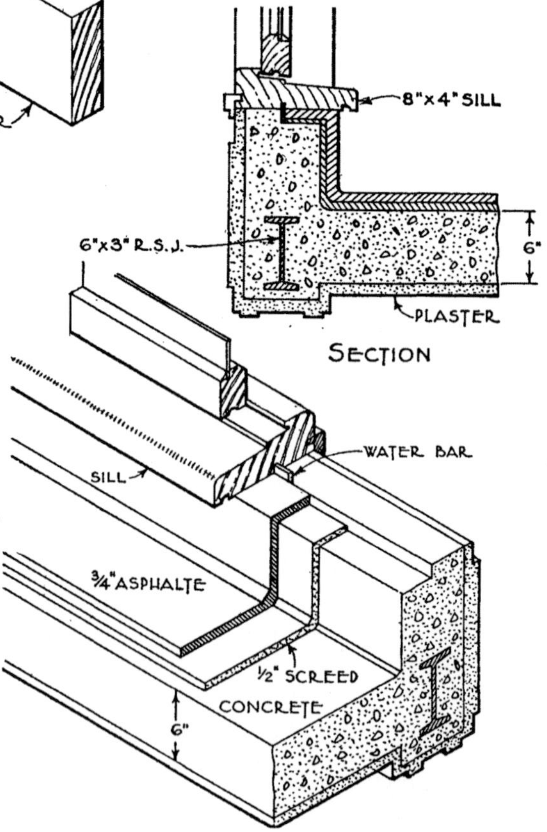

Steel Construction

Fig. 15 shows a lantern light in all-steel construction. The angle shafts are $2\frac{1}{4}$ in. square, and the metal $\frac{3}{16}$ in. thick; the cill is also of the same section. The cill is fixed to the concrete curb by means of a lug, as shown, or by the use of a coachscrew, where the curb is formed with timber.

" Glass-crete "

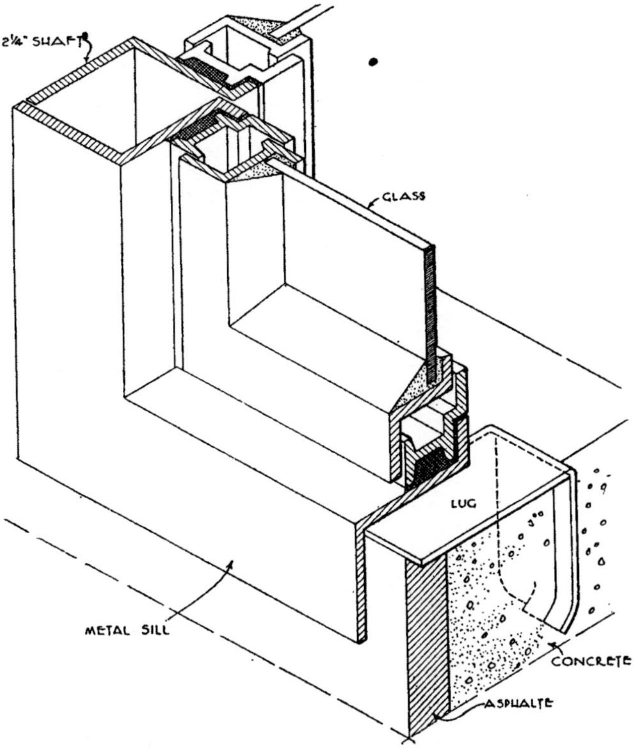

Fig. 15.—Lantern light in all-steel construction

The use of glass prisms embedded in concrete gives a strong construction which allows considerable carrying power over large spans. The glass units are produced in dimensions to conform to structural requirements as to thickness and weight, depending on the type of structure. The use of concrete permits of greater possibilities, and the older methods of glazing, which would not meet the higher standard of modern practice, have been superseded.

All the disadvantages which arose from the use of wrought iron are removed by the use of reinforced concrete. In addition, there are no surfaces to rust,

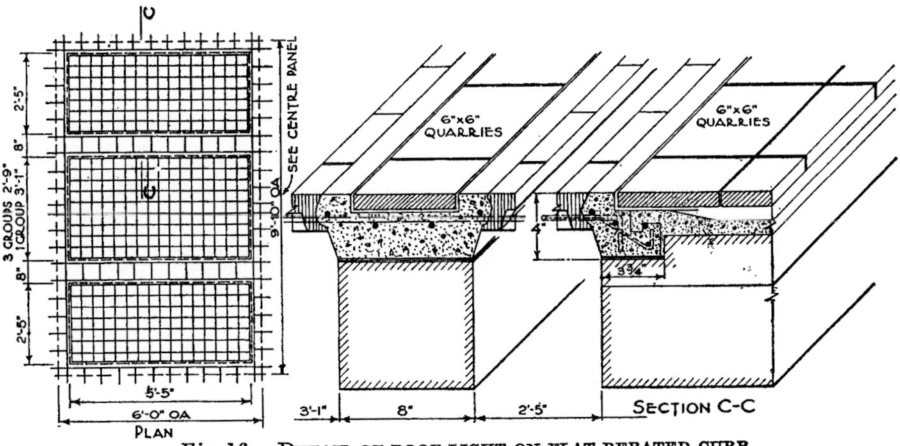

Fig. 16.—Detail of roof light on flat rebated curb

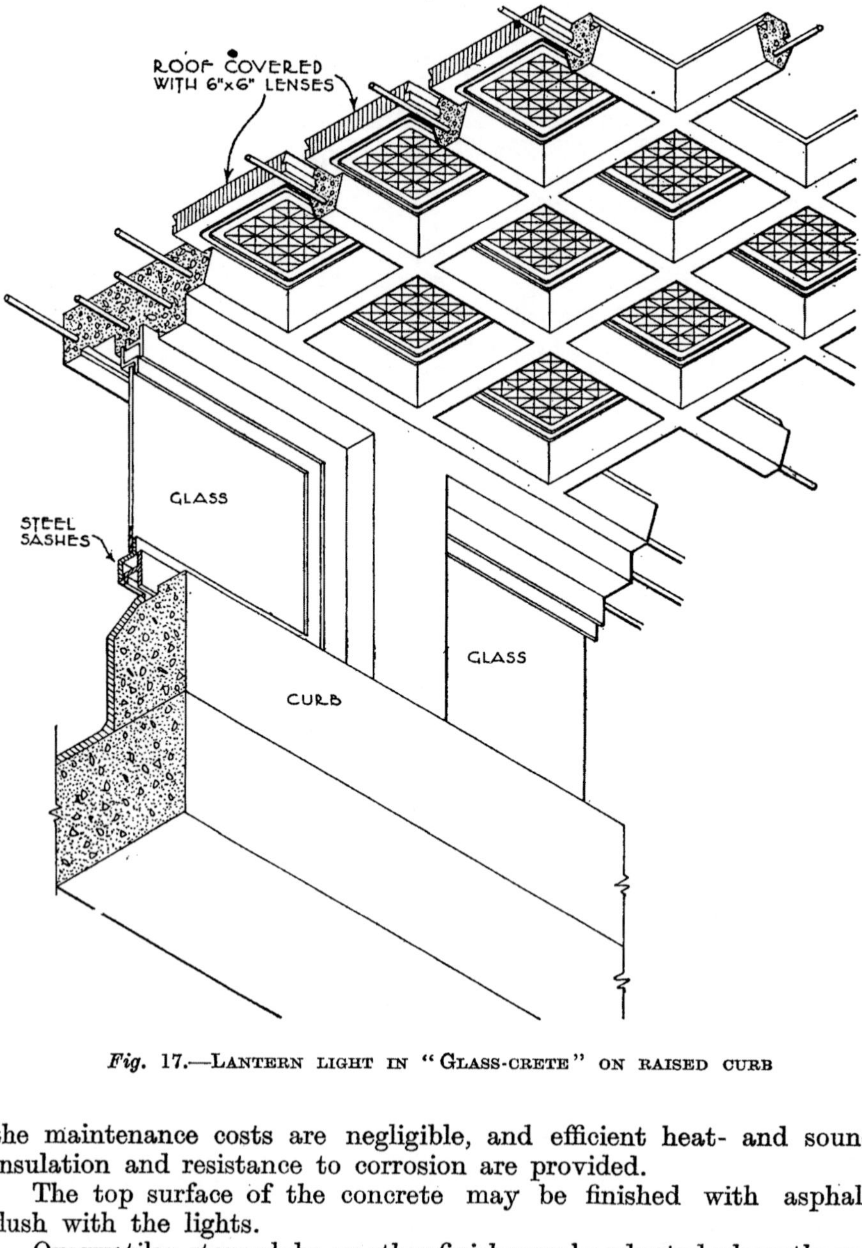

Fig. 17.—LANTERN LIGHT IN "GLASS-CRETE" ON RAISED CURB

the maintenance costs are negligible, and efficient heat- and sound-insulation and resistance to corrosion are provided.

The top surface of the concrete may be finished with asphalte flush with the lights.

Quarry tiles, stone slabs, or other finish may be adopted where the roof light is fixed on a flat rebated curb (see Fig. 16). Fig. 17 shows a lantern light on a raised curb in the same materials. The sides of the lantern are

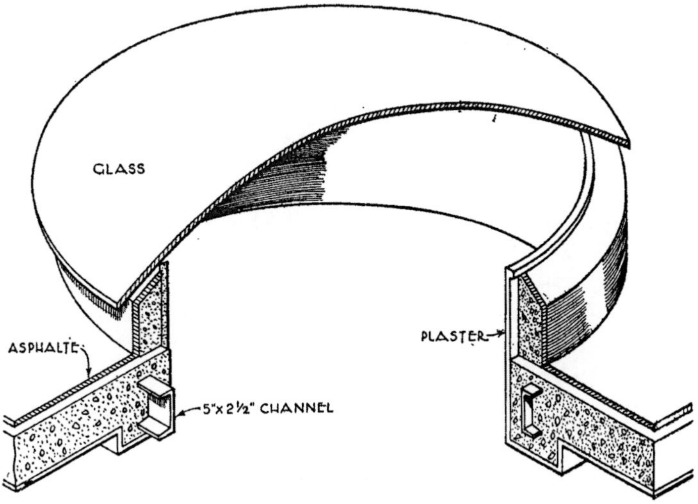

Fig. 18.—Isometric sketch of glass dome light

fitted with steel sashes, and the roof is constructed with 6-in. by 6-in. lenses. Fig. 18 shows a glass dome formed in one piece, and suitable for spans up to 6-ft. diameter. The dome is fixed to the curb by means of special fittings.

Roof Insulation

Flat roofs generally require some form of insulation. Where the roof is of timber construction a certain amount of insulation in some cases is adequate by fixing a wall board to the ceiling. A more even distribution of temperature would be obtained if the insulating board were placed directly under the roof boarding or bitumen sheeting, and using a lath-and-plaster ceiling. When the insulating board is placed directly under the bitumen sheeting or asphalte, it

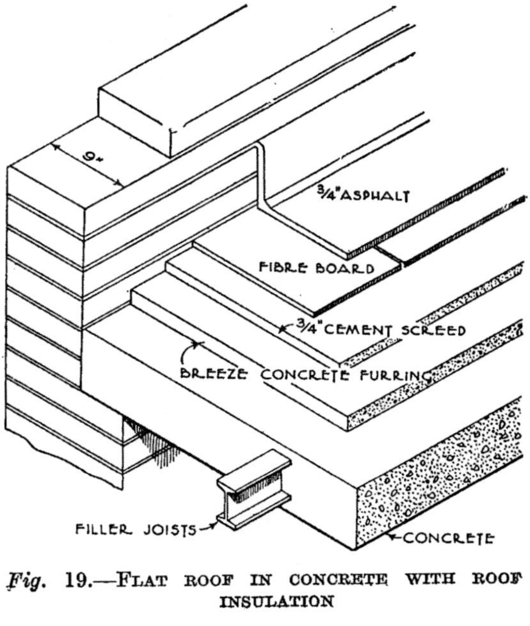

Fig. 19.—Flat roof in concrete with roof insulation

tends to raise the surface temperature during the summer (see Fig. 19). This can be overcome by using asbestos tiles. An excellent tile for this purpose is "Thermotile," which is made with a mixture of asbestos-cement waste. These tiles are laid on three layers of bitumen felt, cemented on to the roof surface and to each other with bitumen. The tiles measure 12 in. by 12 in. by 1 in., and weigh approximately 8 lbs. per square foot. The total thickness, including the waterproof sheeting and bitumen, is approximately 1½ in., and the total weight approximately 10 lbs. per sq. ft. The damp-proofing of the junction between parapet wall and roof is important in this as in other types of flat-roof construction, and details of the method recommended are shown in Fig. 20. Details

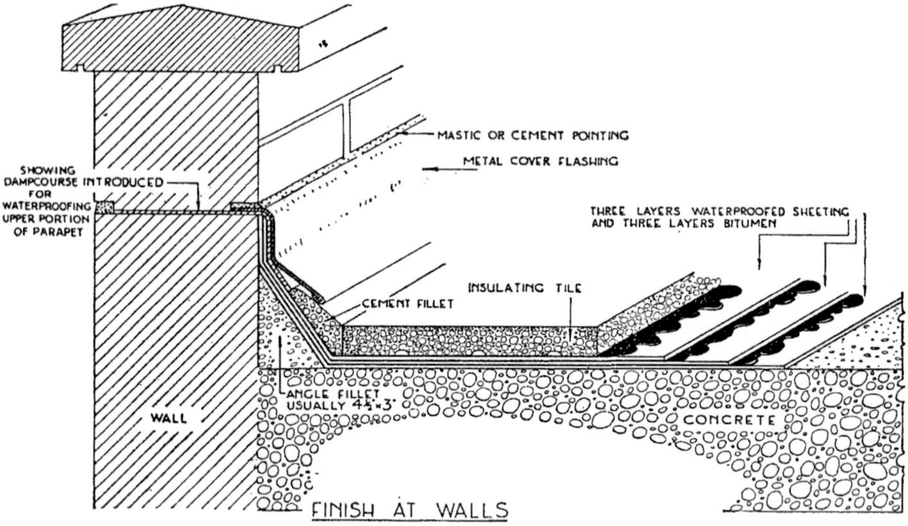

Fig. 20.—Section through concrete roof at parapet

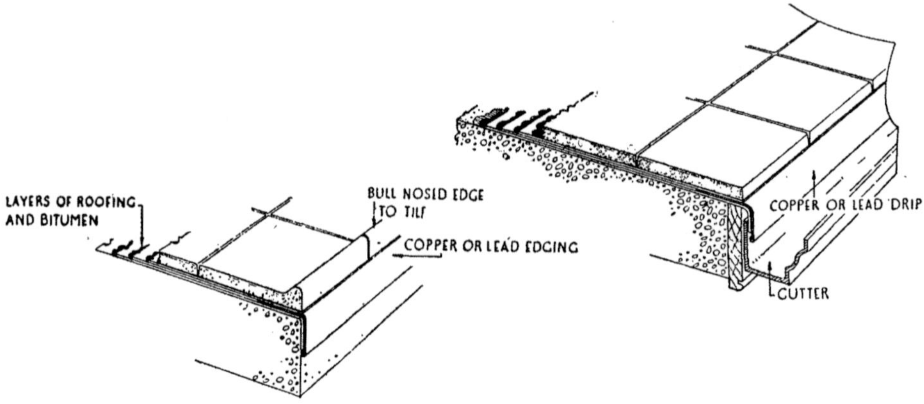

Fig. 21.—Section at verge *Fig.* 22.—Section at eaves

(*D. Anderson & Son Ltd.*)

Fig. 23.—ROOFS COVERED WITH THERMOTILE AT FAIRACRES, ROEHAMPTON LANE, LONDON, S.W. (*D. Anderson & Son Ltd.*)

Architects: Minoprio and Spenceley, A.A.R.I.B.A.
Consulting Architect: B. de Helsby, Esq., A.I.A.A.
General Contractors: C. F. Kearley, Ltd.

Fig. 24.—PAROPA ROOFING, EXTENSION TO MANOR HOUSE HOSPITAL, N.W.
Architects: Bethell and Swannell. (*Frazzi Ltd., London*)

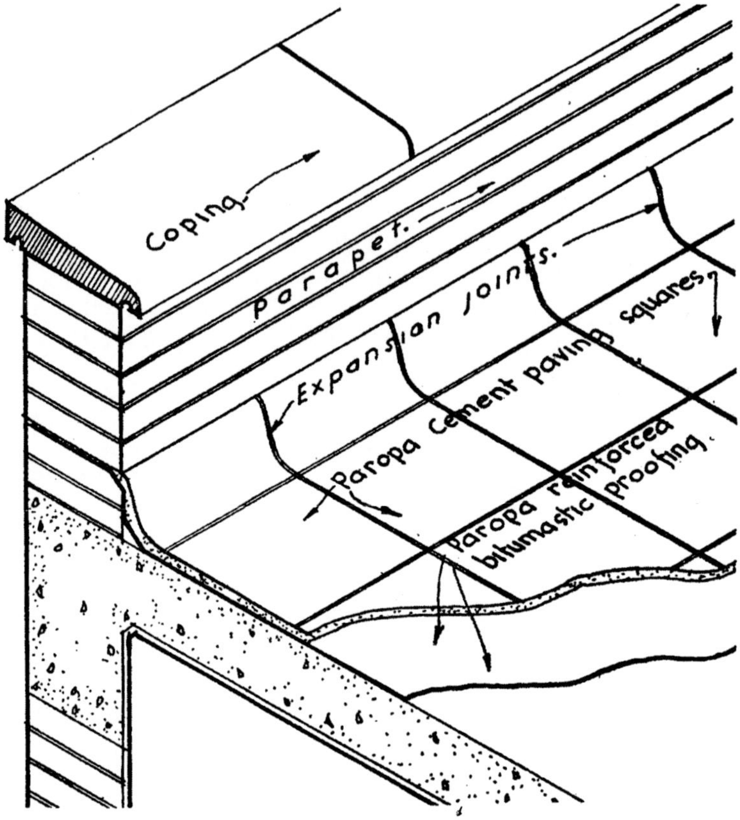

Fig. 25.—PAROPA ROOFING ON CONCRETE, SHOWING TREATMENT AGAINST A BRICK PARAPET. (*Frazzi Ltd.*)

of the treatment at verge and eaves are given in Figs. 21 and 22. The metal cover flashings to walls and eaves are particularly important, and the makers of " Thermotile " roofing, Messrs. D. Anderson & Son Ltd., prefer to execute this work themselves to ensure that it is carried out to exact requirements.

Another excellent type of insulated flat-roof covering is " Paropa," made by Messrs. Frazzi Ltd., which consists of a well-primed surface having a slope of 1 per cent., on which a bituminous mat is built up with layers of pure bitumen, wood felt, and hessian. This waterproof mat is so composed as to ensure complete adhesion to the base. On this impermeable covering is laid a coat of cement mortar of 1 in. thickness, which is well tamped and floated, and while soft is divided into panels, approximately 24 in. square, by means of special tools. The dividing joints framing the panels

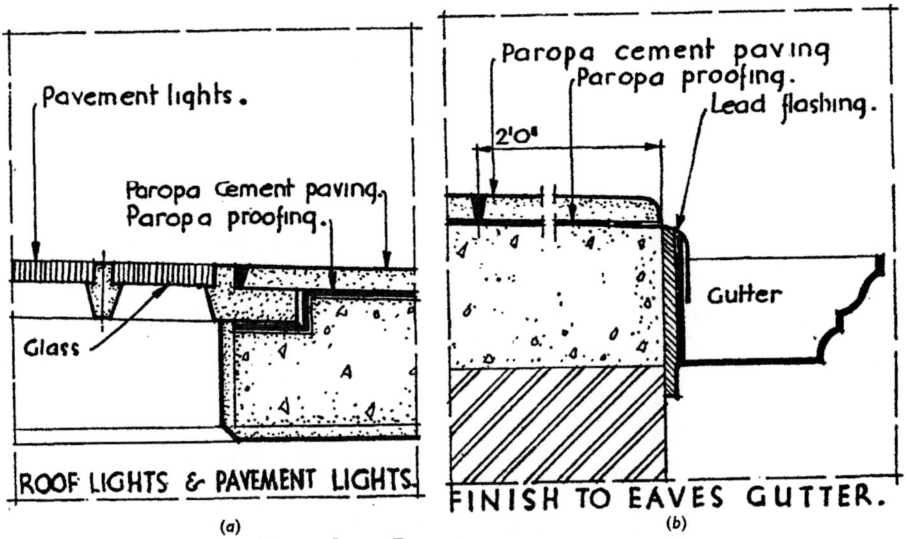

Figs. 26A *and* B.—PAROPA ROOFING ON CONCRETE
a—Junction between Paropa and roof-light. *b*—Showing finish to eaves gutter.
(*Frazzi Ltd.*)

are cut to the bitumen below, thus forming most efficient expansion joints. After hardening of the cement topping the expansion joints are primed and filled with an elastic bitumen.

"Paropa" can be finished in various designs and colours, and forms a good non-slip surface. The weight is 12 lbs. per sq. foot. The junction between roof and parapet, and the treatment at gutters and outlets, together with the treatment of roof lights, are shown in Figs. 25, 26A, B, and C.

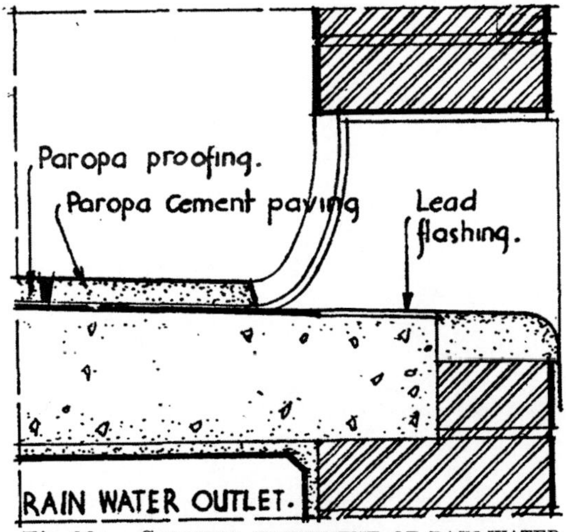

Fig. 26C.—SHOWING TREATMENT OF RAIN-WATER OUTLET IN PAROPA ROOFING ON CONCRETE
(*Frazzi Ltd.*)

Both Thermotile and Paropa can be applied to wood as well as to concrete surfaces, the bitumen layers being applied, preferably, to a layer of insulating board laid on top of the normal roof boarding.

Where a high degree of insulation is required, the use of 2-in. insulating blocks such as "Fosalsil" is a proved method, these being laid on

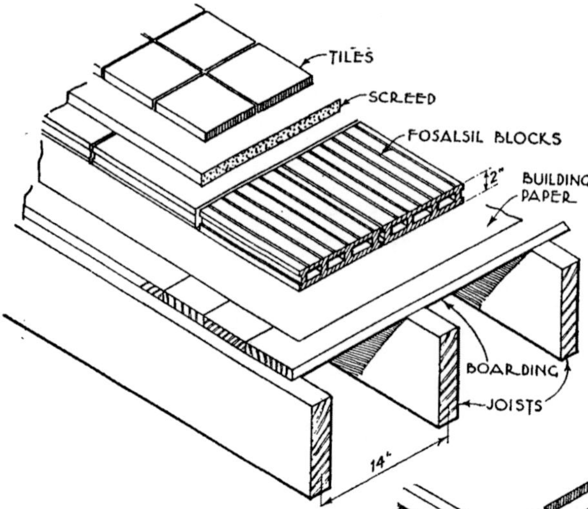

Fig. 27 (*left*).—Roof insulation

Insulating blocks laid on building paper, the base being timber construction.

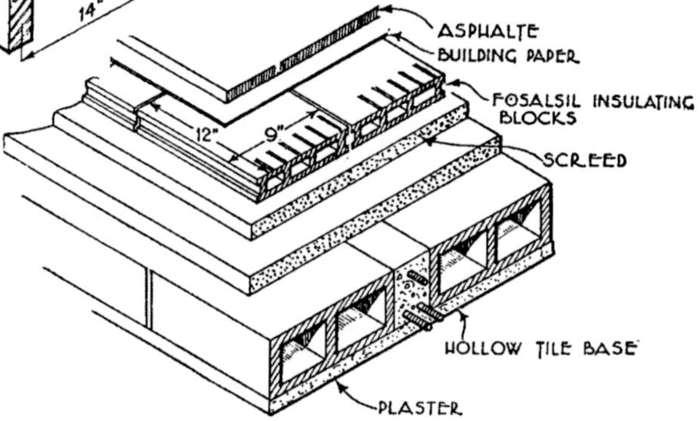

Fig. 28.—Roof insulation

Insulating blocks on hollow-tile base. Methods shown in Figs. 27 and 28 give a high insulation value.

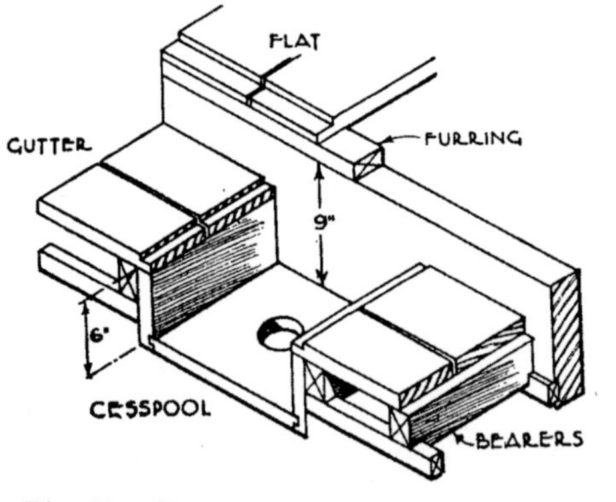

Fig. 29.—Parallel gutter to flat roof

Showing detail of cesspool with joists parallel to gutter.

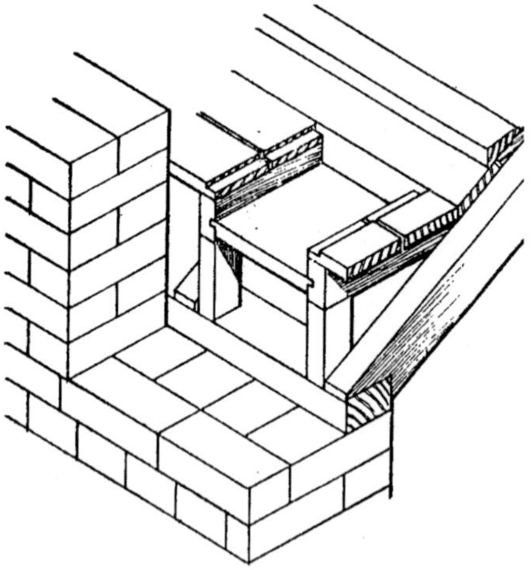

Fig. 30.—Tapering gutter

Isometric sketch of chute to parapet gutter. Note bearers are nailed to side of rafters.

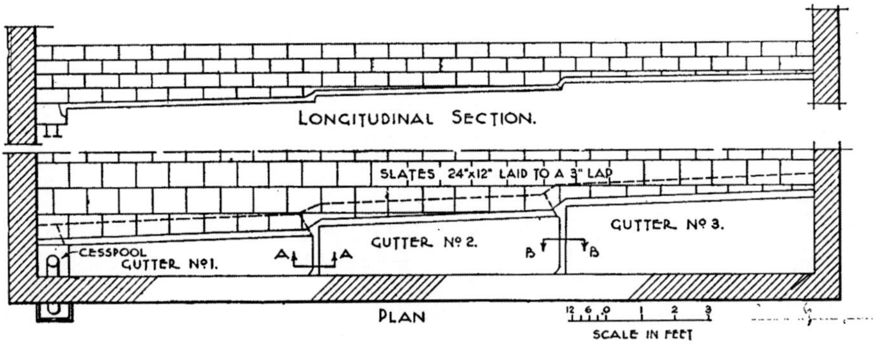

Fig. 31.—Lead gutter to parapet wall

top of boarding or screed for wood or concrete roofs respectively, followed by normal asphalte or paving tile construction. Figs. 27 and 28 illustrate the method for both types of roof.

Preparation for Gutters

Gutters may be either parallel or tapering on plan.

Parallel Gutters

These are formed by fixing a pole-plate at a distance of from 9 in. to 12 in. from the wall. This pole-plate carries the lower ends of the rafters.

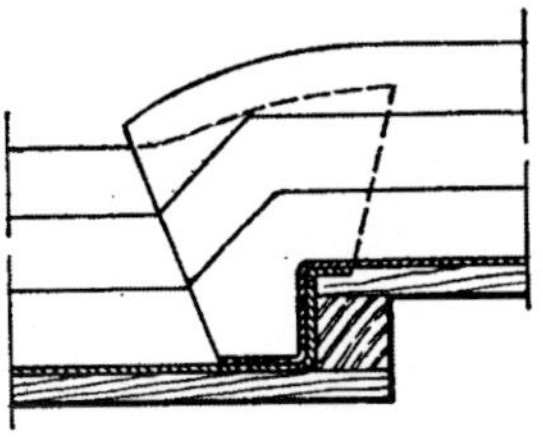

Section through drip A–A from wall side.

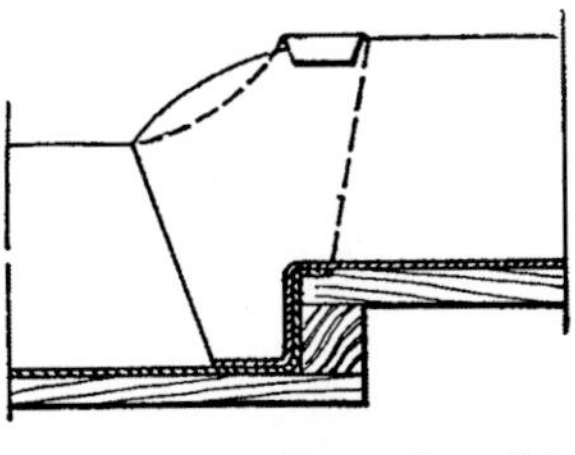

Section through drip showing finish of lead against wall.

Fig. 32

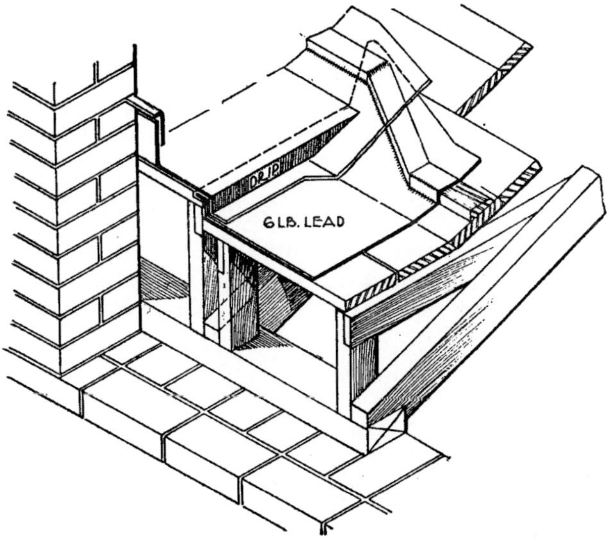

Fig. 33.—Isometric sketch of a tapering gutter, showing drip

Wall removed to show construction more clearly.

Between the wall and the pole-plate short pieces of timber called bearers are fixed, to carry the gutter boarding. These bearers are fixed at 15-in. centres, and must be so fixed as to give the necessary fall and drips when the boards are laid. Fig. 29 shows part of a parallel gutter to a flat roof, together with the construction of the cesspool.

Tapering Gutters

These differ from parallel gutters in being formed on the top of the rafters, so that as the gutter rises from the outlet its width increases. The bearers in these cases are nailed to the side of rafters (see Fig. 30). Fig. 34 shows the method of setting out, and Fig. 31 plan and

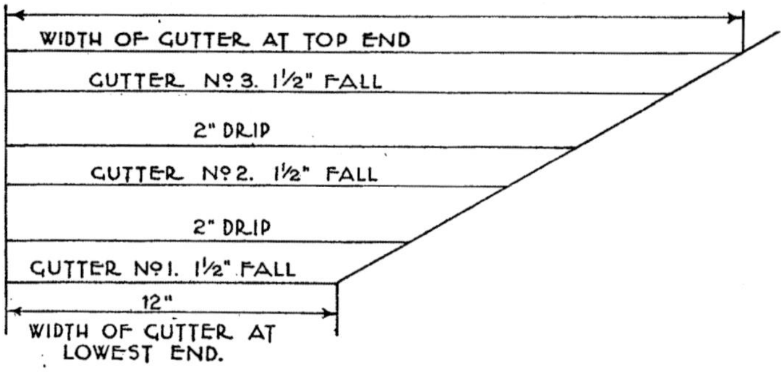

Fig. 34.—DIAGRAM SHOWING WIDTH AT TOP AND BOTTOM OF EACH GUTTER

section of a complete gutter. Fig. 32 shows sections through a drip, one from the wall side and the other from the roof side, and an isometric sketch of the drip is shown in Fig. 33, the wall being removed to show the construction more clearly.

Chute

In some cases, instead of discharging the rain-water into a cesspool, it is taken, through an opening formed in the wall by means of a chute, direct into a rain-water head. The construction is shown in Fig. 31

Chapter IV

SIZES AND SPACING OF ROOF MEMBERS

ALTHOUGH suitable sizes for roof members are given in the tables and illustrations in this and other chapters, everyone interested in building work should be able to calculate the most economical spacing of the rafters, and also the safe load on the beams.

Under the Code of Practice, 1932 (sec. 35), sloping roofs of over 20° pitch may be designed for a combined superimposed and wind load of 15 lbs. per square foot, assumed acting normal to the surface inwards on the windward side, and 10 lbs. per square foot of surface similarly acting outwards on the leeward side, provided this requirement applies only in the design of the roof structure. Snow may be taken at 3 lbs. per sq. ft.

The weight of the truss is given by Fowler's formula : ·04 span + ·4 for light roofs, and ·06 span + ·6 for heavy roofs ; or by Ricker's formula :

$$W = \frac{\text{span}}{25} + \frac{\text{span}^2}{6{,}000}.$$

Working Stresses for Rafters

A list of sizes for purlins is given on page 44.

In calculating the safe load on timber beams, rafters, or joists, various methods are used. The following assumes good-quality timber.

One method is based on the results of numerous tests on small pieces of timber 1 in. by 1 in. in section, supported at both ends, over a span of 12 in., and loaded in the middle, the breaking weight being used as a constant. When the test piece is loaded in this way it is said to be under transverse stress, the fibres at the top being in compression and those at the bottom being in tension, while the intensity of each stress diminishes towards the centre of the beam, or the centre of gravity of the section, at which point they both vanish. Where this occurs it is called the *neutral layer* on the elevation of the beam, and the *neutral axis* in the cross-section. In beams of symmetrical section it is found at the geometrical centre of the section. The unit of transverse stress is called the *modulus of rupture.* It is sometimes supposed to be the same as the maximum fibre stress, but it is found to be greater than this.

The strength varies as $\frac{bd^2}{L}$, where b = breadth in inches, d = depth in inches, and L = span in feet. This formula does not make provision for any variation in the class of material used. This provision is expressed by the constant c, or unit load to break unit beam. It varies with the variety of wood used. For spruce or fir it is found to be 3·5 cwts., for red deal 4 cwts., for oak 4·5 cwts., and for pitch pine 5 cwts.

In the ordinary way the beam is required to carry a load safely, and to do this it is necessary to reduce the load to a fraction of the breaking weight, say $\frac{1}{5}$ or $\frac{1}{6}$, and the factor of safety F is then said to be 5 or 6, as the case may be.

The formula now becomes $\frac{cbd^2}{FL}$.

WEIGHTS OF ROOFING MATERIALS

Material	*Average Weight per Ft. Super in Lbs.*
Slates—North Wales: Firsts	5·5
Seconds	6·5
Thirds	7·5
Cornish: Firsts	6·72
Seconds	7·84
Thirds	11·20
Westmorland: Firsts	8·4
Seconds	10·1
Thirds	12·32
Tiles—Machine-made	11·25
Hand-made	12·25
Pantiles	11·25
Asbestos tiles	3·0
Asbestos corrugated sheets	3·1
Steel corrugated sheets—18 B.W.G.	2·75
20 B.W.G.	2·25
Roof boarding—1 in.	2·5
$\frac{7}{8}$ in.	2·25
$\frac{3}{4}$ in.	1·87
Tiling battens (4-in. gauge)—2 in. by 1 in.	1·5
2 in. by $\frac{3}{4}$ in., or $1\frac{1}{2}$ in. by 1 in.	1·125
1 in. by 1 in.	0·75
Felt.	0·5
	Average Weight per Sq. Ft. in Lbs.
2-in. by 3-in. common rafters at 15-in. centres, 6-ft. span	1·2
2-in. by 4-in. ,, ,, ,, ,, ,, 7-ft. 6-in. span	1·6
2-in. by $4\frac{1}{2}$-in. ,, ,, ,, ,, ,, 9-ft. span	1·8
2-in. by 5-in. ,, ,, ,, ,, ,, 10-ft. ,,	2·0
2-in. by 6-in. ,, ,, ,, ,, ,, 12-ft. ,,	2·4

PURLINS FOR ALL ROOFS

Clear Span	*Distance Apart Not Exceeding*					
	6 *ft.*		7 *ft.* 6 *in.*		9 *ft.*	
	in. in.	*lb./sq. ft.*	*in. in.*	*lb./sq. ft.*	*in. in.*	*lb./sq. ft.*
6	3 × 5	·62	3 × $5\frac{1}{2}$	·55	3 × 6	·50
8	3 × 6	·75	3 × 7	·70	4 × 7	·70
10	4 × 7	1·2	4 × 8	1·1	5 × 8	1·1
12	4 × 9	1·5	$4\frac{1}{2}$ × 9	1·35	5 × 9	1·25
14	5 × 9	1·6	$4\frac{1}{2}$ × 10	1·50	$4\frac{1}{2}$ × 11	1·37

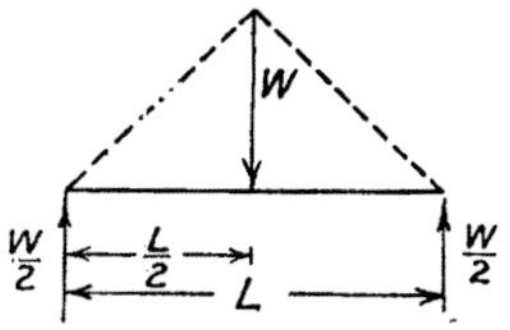

Fig. 1.—BENDING-MOMENT DIAGRAM FOR CONCENTRATED LOAD

Example.—Required the safe concentrated load in cwts. on a beam of red deal 5 in. deep, 2 in. wide, and 10-ft. span. F = 5.

$$W = \frac{cbd^2}{FL} = \frac{4 \times 2 \times 5^2}{5 \times 10} = 4 \text{ cwts.}$$

Resistance Moment

As a means of comparison we will now ascertain the load based on the moment of resistance of the beam. This is given by the following equations:—

$$BM = MR = Zf, \text{ in which—}$$

BM = bending moment.
MR = moment of resistance.
Z = section modulus of the beam.
f = permissible fibre stress in the material, usually referred to as the extreme fibre stress.

The bending produced by the load on a simply supported beam tends to shorten or compress the upper fibres, and to lengthen or put in tension the lower fibres. In the case of a simply supported beam, with a load at the centre, the maximum bending moment is $\frac{WL}{4}$, and is obtained by multiplying the reaction $\frac{W}{2}$ by the leverage $\frac{L}{2}$ (see Fig. 1). The modulus of section, $Z = \frac{bd^2}{6}$, is obtained as follows: on the section shown in Fig. 2 draw two diagonals, forming triangles, the two middle triangles showing the quantity and distribution of stress. Rectangular sections are assumed as if they were two triangles joined together at their apexes, and in consequence equally stressed in every fibre. The area ACE will be the equivalent area for compression, and EBD the equivalent area for tension. The area of each hypothetical triangle would be $\frac{b}{2} \times \frac{d}{2} = \frac{bd}{4}$.

Given that the centre of gravity of every triangle is at one-third the height from the base, and the height of the triangle is $\frac{d}{2}$, the leverage will be $\frac{d}{2} \times \frac{2}{3} = \frac{d}{3}$, and for the two triangles $\frac{2}{3}d$. The section modulus Z for any rectangular section will be the area of one triangle multiplied by the lever arm, $= \frac{bd}{4} \times \frac{2}{3}d = \frac{bd^2}{6}$.

For the specimen under consideration, the value of $Z = \frac{1 \times 1^2}{6} = 0{\cdot}166 \text{ in.}^3$. Since the specimen broke at 4 cwt., and the factor of safety

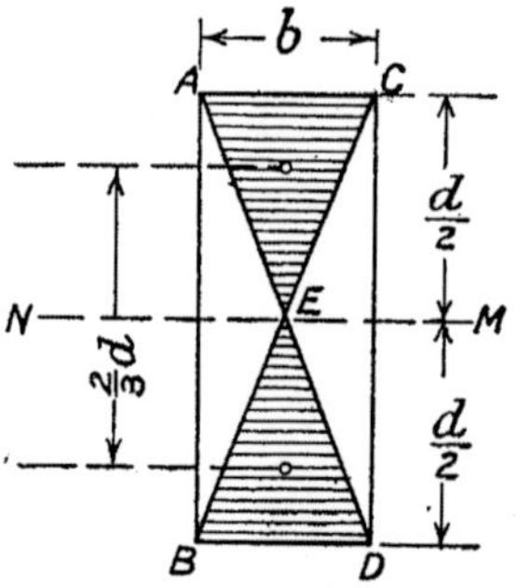

Fig. 2.—DIAGRAM SHOWING RESISTANCE AREA

is 5, the safe load will be 0·8 cwt., which gives a bending moment of—

$$BM = \frac{WL}{4} = \frac{0{\cdot}8 \times 112 \times 12}{4} = 268{\cdot}8 \text{ lb. in.,}$$

$$\text{therefore } f = \frac{BM}{Z} = \frac{268{\cdot}8}{0{\cdot}166} = 1{,}619 \text{ lb.}$$

Now to complete the problem :—

$$\text{If } \frac{WL}{4} = Zf, \text{ then } W = \frac{4Zf}{L}.$$

$$W = 4\,\frac{bd^2 \times f}{6 \times L \times 12} = \frac{4 \times 2 \times 5^2 \times 1619}{6 \times 10 \times 12}$$

$= 448$ lbs. or 4 cwts.

The value of f_0 obtained by experiment is called the *modulus of rupture*, and will be found to be 18 times the weight that will break the beam across. This will be seen by substituting the values of *l*, *b*, and *d* for the experimental beam in the equation.

$$f_0 = \frac{WL}{4} \times \frac{6}{bd^2} = \frac{W \times 12 \times 6}{4 \times 1 \times 1^2} = 18W$$

The value of f obtained by the experiments above referred to is in most cases too high, because the specimens selected for tests are usually small pieces.

WORKING UNIT STRESSES FOR STRUCTURAL TIMBER IN POUNDS PER SQUARE INCH

Timber	*Extreme Fibre Stress*		*Modulus of Elasticity, Average*	*Shearing Stress*	
	Average Ultimate	*Safe Stress*		*Parallel to Grain*	*Bending*
Spruce	5,000	1,000	1,300,000	100	80
Redwood	5,600	1,100	1,200,000	150	100
Douglas fir	6,100	1,200	1,510,000	180	120
Pitch pine	6,500	1,400	1,600,000	180	120
Oak	6,500	1,400	1,200,000	200	120

Design of Boarding

In making up the load carried by the various members it will be found convenient to determine the resultant load carried by one square foot. From the Table previously given, slates (thirds) weigh 7·5 lbs., battens (2 in. by ¾ in.) 1·125 lbs., felt 0·5 lb., and boarding 2·5 lbs., per square foot respectively. The dead load will then be 11·625 lbs. per square foot of roof area. To this must be added half the proposed snow load (1·5 lbs.), as a maximum wind pressure (15 lbs.) and a snow load are not likely to occur at the same time. Adding these amounts together the total load including the boarding will be 28·125 lbs. per square foot.

Assuming that the rafters are spaced at 15-in. centres, the moment due to the load will be $\frac{1}{8}$ $Wl^2 = \frac{28{\cdot}2 \times 1{\cdot}25^2 \times 12}{8} = 66$ in. lbs. approximately.

This moment is resisted by a $\frac{7}{8}$-in. by 12-in. section of boarding, for which the resulting fibre stress will be $f = \frac{BM \times 6}{bd^2} = \frac{66 \times 6}{12 \times \frac{7}{8}^2} = 43$ lbs. per square inch, which is very low, and indicates that for ordinary conditions the design for boarding need not be carried out.

Design of Rafters

For the rafters assume a 2-in. by 4-in. section, and 8-ft. span. At 2 lbs. per foot run the dead weight per foot of rafter is $\frac{2 \times 8}{8 \times 1{\cdot}25} = 1{\cdot}6$ lbs. Adding the weight of the rafter the total load to be carried by the rafter is a uniform load of 29·8 lbs. per square foot.

The moment is $\frac{Wl^2}{8} = \frac{29{\cdot}8 \times 8 \times 1{\cdot}25 \times 8 \times 12}{8} = 3{,}576$ lbs. in.

The section modulus of a 2-in. by 4-in. rectangle is $\frac{1}{6}bd^2 = \frac{1}{6} \times 2 \times 4^2$ $= 5{\cdot}33$ in.^{3}, and the fibre stress is $f = \frac{BM \times 6}{bd^2} = \frac{3576 \times 6}{2 \times 4^2} = 670{\cdot}5$ lbs. per square inch. As the allowable fibre stress is 1,000 lbs. per square inch, the assumed section is sufficient.

The maximum span for the rafters to carry the above loading of boarding, battens, and slates, and also a wind pressure of 15 lbs. per square foot, making a total of 29·8 lbs. per square foot, may be worked out as follows :—

Assuming a safe fibre stress of 1,000 lbs. per square inch, and a nominal breadth b of the rafters as 2 in., spaced at 15-in. centres :—

Then, $\quad BM = \frac{WL}{8}, \quad W = L \times \frac{15}{12} \times 29{\cdot}8$ lbs. $= 37{\cdot}25L$.

Therefore $BM = \frac{37{\cdot}25L \times L \times 12}{8} = 55{\cdot}87L^2$.

The moment of resistance $MR = \frac{bd^2}{6} \times f = \frac{2 \times d^2 \times 1000}{6} = \frac{1000d^2}{3}$.

Equating the bending moment to the moment of resistance :—

$$BM = MR$$

$$55{\cdot}87L^2 = \frac{1{,}000d^2}{3}$$

$$\frac{55{\cdot}87L^2 \times 3}{1000} = d$$

$$d = \sqrt{\cdot 1676L} = \cdot 409L.$$

The depth of the rafters will therefore be, say, ·41 of the span in feet. For $f = 1{,}100$ lbs., $d = \cdot 39L$.

Where hand-made tiles are used in lieu of slates, the loading will be increased by the difference between the weight of the slates and the weight of the tiles, i.e. the difference between 7·5 and 12·3 lbs., or 4·8 lbs., making a total load of 34·6 lbs. Working this out in the same way as that used for slates, the depth will be found to be ·441L for $f = 1{,}000$ lbs., and ·42L for $f = 1{,}100$ lbs. These rules are shown plotted in the graph, Fig. 3.

When the rafters are required to carry a plaster ceiling, the deflection is usually limited to $\frac{1}{360}$ of the span to minimise cracks in the plaster. This condition will necessitate the rafters being designed for stiffness instead of strength. Taking the safe fibre stress at 1,000 lbs. per square inch, and E the modulus of elasticity at 1,510,000 :—

The deflection of a rectangular beam under a uniform load is given by the formula $\Delta = \frac{5}{384}\frac{WL^3}{EI}$, or for a maximum deflection of $\frac{1}{360}$:—

$$\frac{5}{384}\frac{WL^3}{EI} = \frac{L}{360}.$$

Dividing by L and substituting the given value of E, we get :—

$$\frac{5WL^2}{384 \times 1{,}510{,}000\,I} = \frac{1}{360}$$

$$\frac{WL^2}{322133} = I \qquad (1)$$

Again, for a beam of any symmetrical section, we know that :—

$$\frac{\text{Moment of inertia of section}}{\text{Modulus of section}} = \frac{\text{depth}}{2}.$$

Now $Z = \text{moment of resistance} \div \text{stress} = \frac{MR}{f}$. Hence, substituting for Z :—

$$I = \frac{MR}{f} \times \frac{D}{2}.$$

But for a beam with a distributed load, $BM = \frac{WL}{8}$,

$$I = \frac{WL}{8f} \times \frac{D}{2} = \frac{WLD}{16f} \qquad (2)$$

Combining results (1) and (2) we have :—

$$\frac{WL^2}{322{,}133} = \frac{WLD}{16f}$$

$$\frac{L}{D} = \frac{322{,}133}{16 \times 1000} = \frac{20}{1}.$$

Hence the depth of a uniform beam must not be less than $\frac{1}{20}$ of the span in feet, or $\frac{1 \times 12}{20} = \cdot 6$ of the span in inches if the deflection has not to exceed $\frac{1}{360}$ of the span (see graph, Fig. 3).

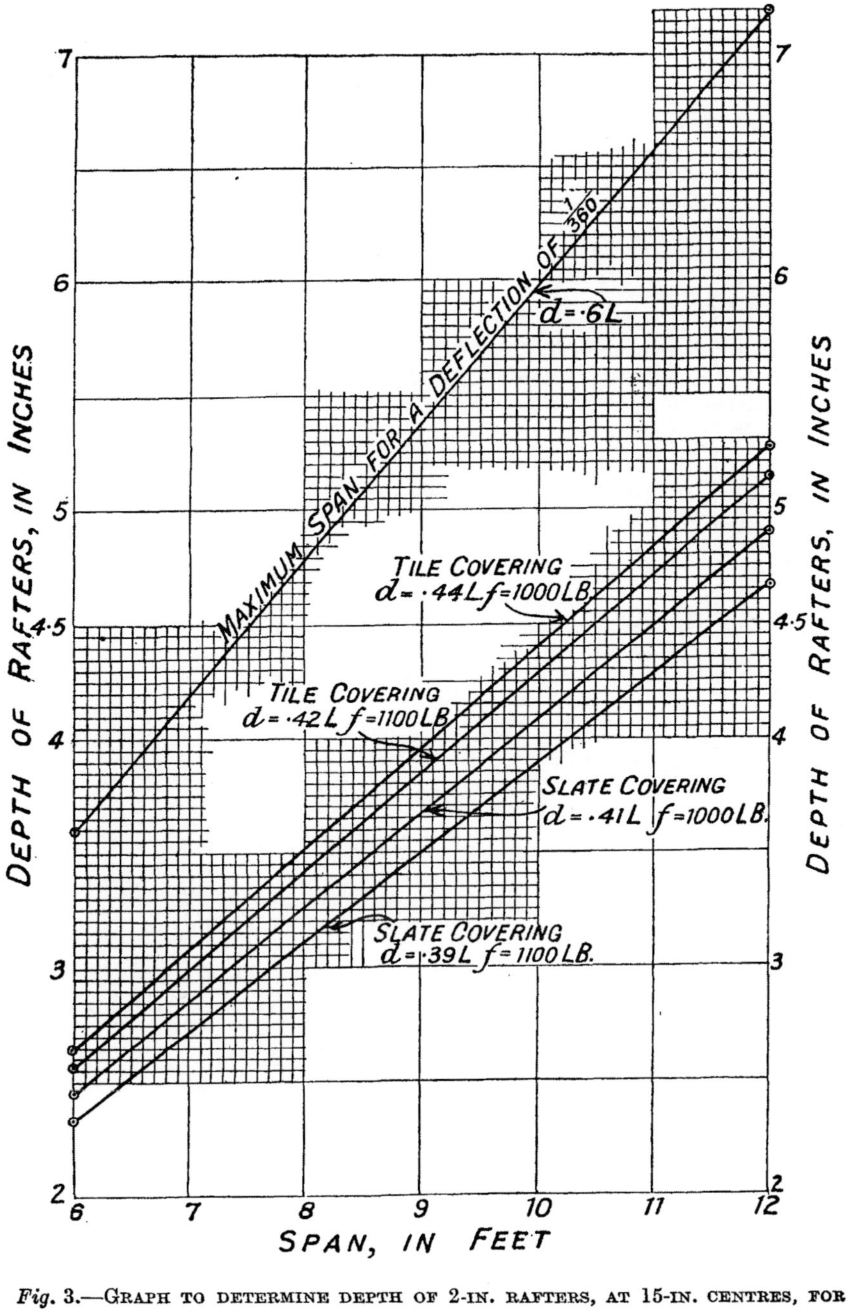

Fig. 3.—Graph to determine depth of 2-in. rafters, at 15-in. centres, for slates and tiles

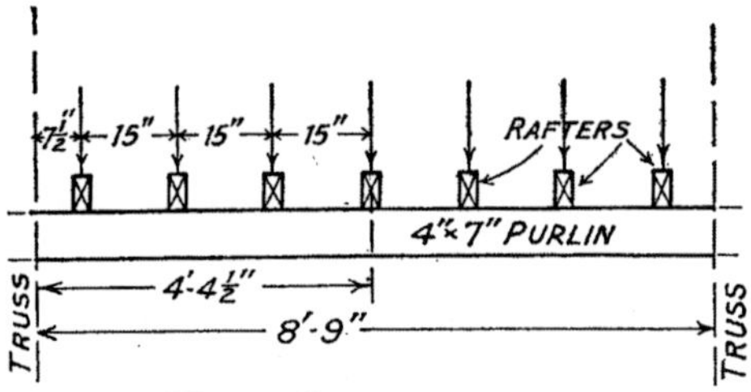

Fig. 4.—LOADS ON PURLIN

Maximum Span

The maximum span for rafters of a nominal breadth of 2 in. and a spacing of 15 in., to carry slates or tiles on boarding, etc., is given in the Table below.

Design of Purlins

The purlin section is set at right angles to the rafter. It is then subjected to a normal load due to the rafters. The loads coming on the purlins from the rafters may be considered as being distributed, which gives a maximum moment slightly less than when they are considered as concentrated loads on each rafter (see Fig. 4). The span will be taken as 8 ft. 9 in., and the distance apart 8 ft., giving an area of 70 sq. ft. The weight of the purlin over this area is ·86 lb. per square foot, therefore the weight to be designed for is 29·8 + ·86 = 30·66 lbs. per square foot over an area of 70 ft., = 2,146·2 lbs.

Common Rafters Nominal Dimensions	*Maximum Span*	
	Slates	*Tiles*
in. in.	*ft.*	*ft.*
2 × 3	7·3	6·8
2 × 4	9·75	9·1
2 × 4½	11·0	10·2
2 × 5	12·1	11·3
2 × 6	14·6	13·4

Using a 4-in. by 7-in. section as given in the Table on page 44, the safe fibre stress will be :—

$$f = \frac{2146 \times 8{\cdot}75 \times 12 \times 6}{8 \times 4 \times 7 \times 7} = 860 \text{ lbs. per square inch.}$$

Taking the loads to act at the centre of the rafters, the maximum moment is :—

$$M = \left[\left(\frac{2146}{2} \times \frac{8{\cdot}75}{2}\right) - \frac{2146}{7}(1{\cdot}25 + 2{\cdot}5 + 3{\cdot}75)\right]12$$

$$= [(4694{\cdot}375) - (2299{\cdot}125)]12 = 28{,}743 \text{ in. lbs.}$$

$$\text{Modulus section of purlin} = \frac{4 \times 7 \times 7}{6} = 32{\cdot}66.$$

$$\text{Maximum stress in the timber} = \frac{28743}{32{\cdot}66} = 880 \text{ lbs.}$$

Stress Diagrams for Roof Trusses

The method by which the stress in each member of a roof truss is found graphically is explained in the following example of a king-post roof truss. Suppose the span to be 24 ft., trusses placed at 10-ft. centres, the roof slope 30°, and the covering, slates on battens and boarding (see Fig. 5).

Assume weight or dead load as previously worked out is 15·66 lbs. per square foot, but to this must be added the weight of the truss. Using Jacoby's formula for wooden roofs, W = ·075 span + ·5 = 24 × ·075 + ·5 = 2·3 lbs. per square foot, making a total load of 17·96 lbs. per square foot. The wind pressure of 15 lbs. per square foot of surface is assumed to be acting normal to the surface inwards on windward side, and 10 lbs. per square foot of surface similarly acting outwards on the leeward side, and a ceiling load of 10 lbs. per square foot.

The slope length is $\dfrac{\text{half-span}}{\cos\ 30^\circ} = \dfrac{12}{0{\cdot}866} = 13{\cdot}85$ ft.

Then *dead load* $= 13{\cdot}85 \times 10 \times 17{\cdot}96 = 2{,}487{\cdot}4$ lbs.

Load on central points $= \dfrac{2487{\cdot}4}{2 \times 112} = 11{\cdot}1$ cwts.

Load on end points $= \dfrac{2487{\cdot}4}{4 \times 112} = 5{\cdot}5$ cwts.

Wind load $= 13{\cdot}85 \times 10 \times 15 = 2{,}077{\cdot}5$ lbs.

Load on central point $= \dfrac{2077{\cdot}5}{2 \times 112} = 9{\cdot}27$ cwts.

Load on end points $= \dfrac{2077{\cdot}5}{4 \times 112} = 4{\cdot}63$ cwts.

Suction $= 13{\cdot}85 \times 10 \times 10 = 1{,}385$ lbs.

Load on central point $= \dfrac{1385}{2 \times 112} = 6{\cdot}2$ cwts.

Load on end points $= \dfrac{1385}{4 \times 112} = 3{\cdot}1$ cwts.

Ceiling load $= 24 \times 10 \times 10 = 2{,}400$ lbs.

Load on central point $= \dfrac{2400}{2 \times 112} = 10{\cdot}7$ cwts.

Load on end points $= \dfrac{2400}{4 \times 112} = 5{\cdot}35$ cwts.

The latter are taken directly by the walls, and need not be considered.

To a scale of $\frac{1}{4}$ in. to a foot, draw the frame diagram, and indicate on the figure the forces combining the wind and dead loads. The dead load and wind pressure acting on the same point can be represented in magnitude and direction by two adjacent sides of a parallelogram, and their resultant is represented by the diagonal of the parallelogram

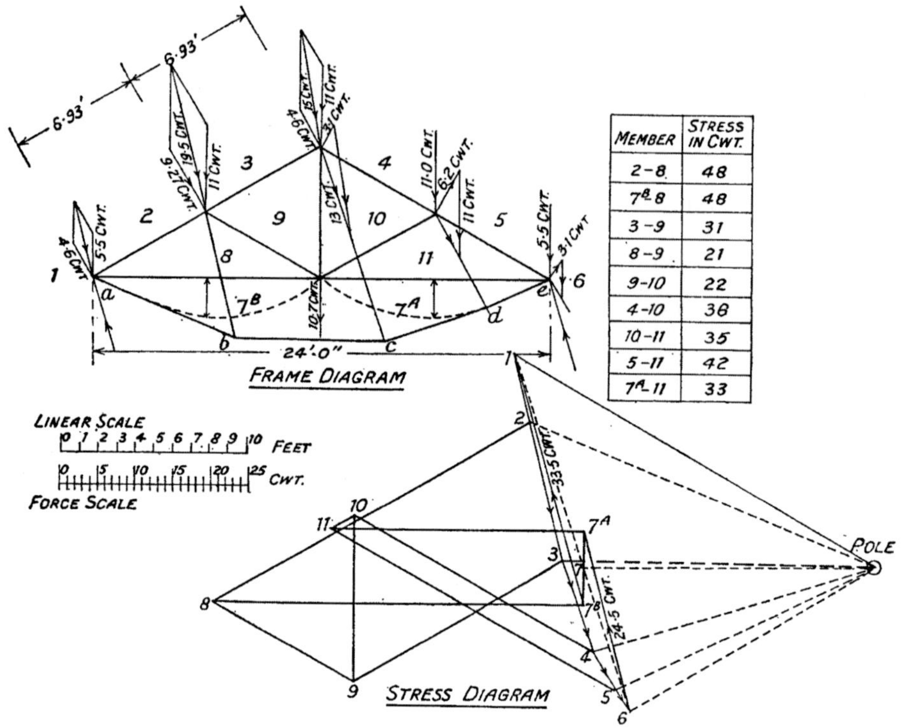

MEMBER	STRESS IN CWT.
2–8	48
7^B–8	48
3–9	31
8–9	21
9–10	22
4–10	38
10–11	35
5–11	42
7^A–11	33

Fig. 5.—FRAME DIAGRAM OF KING-POST WITH STRUCTURAL LOAD, WIND PRESSURE, AND SUCTION

drawn from the point of application. The spaces must next be numbered or lettered between the forces acting on the frame.

Draw the load line 1, 2, 3, 4, 5, 6, and take any pole O. Draw the radial lines O1, O2, etc., and across the corresponding spaces 2, 3, 4, etc., on the frame diagram, draw parallel lines forming the funicular polygon *abcde*. Draw the closing line *ae*, and from the pole O draw a parallel line cutting the load at point 7. If there had been no ceiling load the lines drawn from this intersection to the points 1 and 6 would have given the magnitude and direction of the reactions 1–7 and 6–7. The addition of the ceiling load will obviously increase the reactions, and also cause the reactions to assume a more vertical direction. From the point 7 on the temporary reaction line, set up on each side half the magnitude of the central ceiling load, giving points 7^A and 7^B. Join point 1 to 7^B, and 6 to 7^A This gives the magnitude and direction of the two reactions (see Fig. 5). Next draw the stress lines parallel to the members of the truss : 2–8, 7–8 ; 3–9, 8–9 ; 9–10, 4–10 ; 10–11, 7–11 ; and 5–11. The forces in the frame can now be scaled off, and the magnitude of the stress in each member of the truss tabulated as below :—

Member	2–8	7^B–8	3–9	8–9	9–10	4–10	10–11	7^A–11	5–11
Magnitude of stress, in cwts.	48	48	31	21	22	36	3·5	42	38

The usual formula for roof members in compression is the Rankine-Gordon formula, and is :—

$$P = \frac{fA}{1 + \frac{1}{ac}\frac{l^2}{r^2}}$$

Where P = ultimate load in pounds

$f = \frac{2}{3}$ of the ultimate compressive strength of the material in pounds per square inch, say 4,000 lbs. for fir.

a = constant = 1 for column with round ends; 2 for column with one end rounded.

c = constant = 750 for timber.

l = length of column in inches.

r = least radius of gyration of cross-section $= \sqrt{\frac{I}{A}}$ or $r^2 = \frac{I}{A}$

where I = moment of inertia $= \frac{bd^3}{12}$ for a rectangular section, and A = area of cross-section.

For the principal rafter the maximum stress is 5,376 lbs. and the unsupported length 6·94 ft.

Assume a scantling of 5 in. by 4 in., then :—

$$P = \frac{4000 \times 20}{1 + \frac{1}{2 \times 750}\frac{(6{\cdot}94 \times 12)^2}{1{\cdot}33}} = 17{,}937 \text{ lbs.}$$

Adopting a factor of safety of 4, the safe load will be $\frac{17937}{4} = 4{,}480$ lbs. or 40 cwts. The section 5 in. by 4 in. is too small. Try a 5-in. by 5-in. section and work out as before.

For the struts with a maximum compression of 21 cwts., and an unsupported length of 6·94 ft., try a section 5 in. by 2½ in., then :—

$$P = \frac{fA}{1 + \frac{1}{ac}\frac{l^2}{r^2}} = \frac{4000 \times 12{\cdot}5}{1 + \frac{1}{2 \times 750}\frac{(6{\cdot}94 \times 12)^2}{1{\cdot}33}}$$

$$P = 11{,}185 \text{ lbs.}$$

and $\frac{11185}{4} = 2{,}796$ lbs. or 24 cwts. safe load.

The king-post has a tensional stress of 22 cwts., and allowing 2·5 cwts. per square inch on the net area in tension, the king-post will require an area of $\frac{24}{2\cdot5} = 9\cdot6$ sq. in., say, 4 in. by 3 in.

The tie-beam has a maximum tension of 48 cwts., and a maximum bending moment $\frac{WL}{8} = \frac{10\cdot7 \times 12}{8} = 16$ cwts. ft. or 192 cwts. in.

Try a 10-in. by 5-in. section, then :—

$$-\frac{W}{A} \pm \frac{M}{Z} = -\frac{48}{50} \pm \frac{192}{\frac{1}{6}(5 \times 10^2)}$$

$$= -\cdot96 \pm 2\cdot3 = 3\cdot26 \text{ cwts. per square inch in tension,}$$

and 1·34 cwts. per square inch in compression, which is satisfactory.

OBTAINING ROOF BEVELS

To obtain the roof bevels for a span roof first draw a plan of the roof, as shown in Fig. 1. The bevels for the members may then be obtained as follows :—

Common Rafter

The bevels and lengths may be obtained direct from the section FLH, FH being the rise or height of the roof (Fig. 6).

Hip Rafter

At right angles to the plan of the hip, as AE or ED, set out height of ridge E_1 E_1, that is, the top point of the rafter. Join E_1A or E_1D : then the bevels are obtained as shown. For backing the hip, draw the line 1—2 at right angles to the plan of the hip, and at the point of intersection of these two lines draw the line 3—4 at right angles to AE_1. Rotate 3—4 down to 3—5, and join 1—5, 5—2, and the backing is obtained. A similar construction gives the length, bevels, and backing for the adjacent hip, which is, of course, different owing to the irregular shape of the hipped end.

Jack Rafter

From the true length of the common rafter LH, turn over line to position LF_1. Next produce line of jack rafter as shown at GG_{11}, then top bevel is at G_{11}. The side bevel will be as for the common rafter.

Roof Boarding

Find development of side and end slopes by turning F_1 to F_{11} and drawing $F_{11}E_{11}$ parallel to CD. A complete roof plan is shown in Fig. 7.

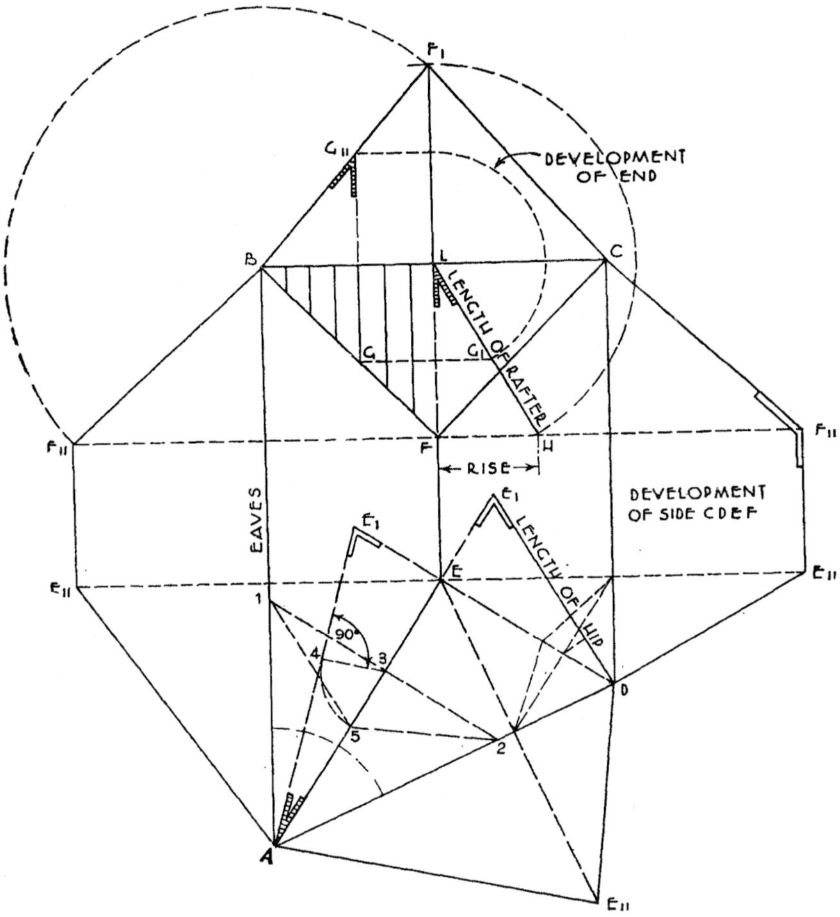

Fig. 6.—THIS ILLUSTRATES THE METHOD OF OBTAINING ROOF BEVELS

A plan of the roof is first drawn. The bevels of the members may then be obtained as described in the text.

Purlins

The bevels for the purlins are shown in Fig. 8. Set up the pitch of the roof, in this case 30°, and draw the plan of the hip at 45°. Place the purlin in position, and letter the points ABC. With B as centre, and radius BA, turn A into a horizontal position A^1. Next drop verticals from A, A^1 on to the plan of the hip as shown, and join A^1A^{11} horizontally. The line drawn through B^1A^{11} gives the top cut for the purlin. The side is obtained in a similar manner.

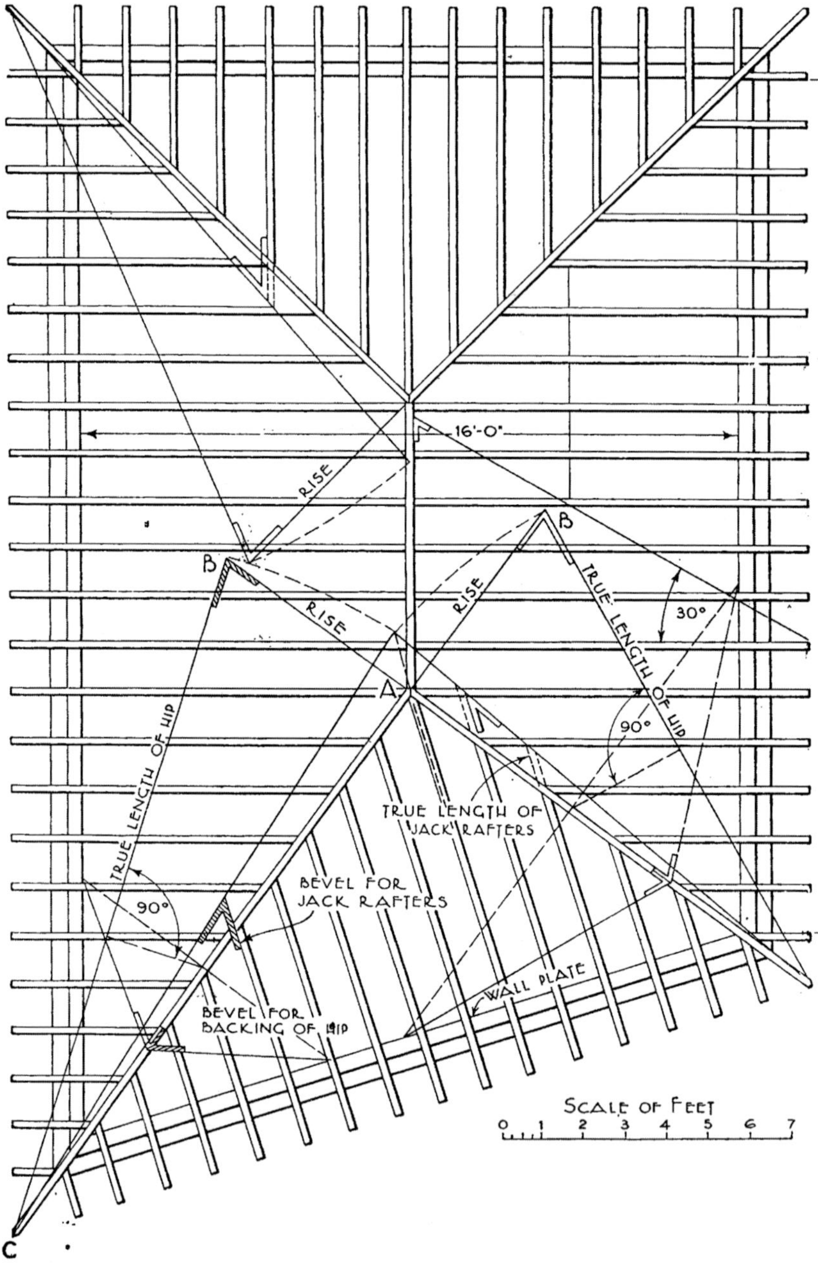

Fig. 7.—PLAN OF ROOF WITH IRREGULAR HIP END

Where Wall Plates Are at Obtuse or Acute Angles

Where the wall plates are at obtuse or acute angles, and the hips do not bisect the angle between them, the finding of the bevels will be slightly different, though based on the same geometrical principles (see Fig. 9). Draw lines ABC to represent the lines of the wall plates, and BD the centre line of the hip. Fix upon any point in it and draw IE, IC parallel to BC, AB respectively. At any convenient point E draw E2 at right angles to BC, and (2) draw 2–3 at right angles to E2. From E set up the pitch of the roof, and 2—4 at right angles to 2C.

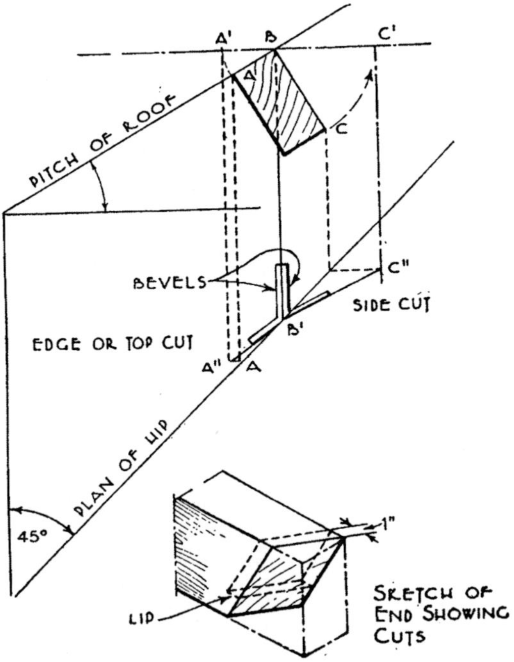

Fig. 8.—Bevels for purlins

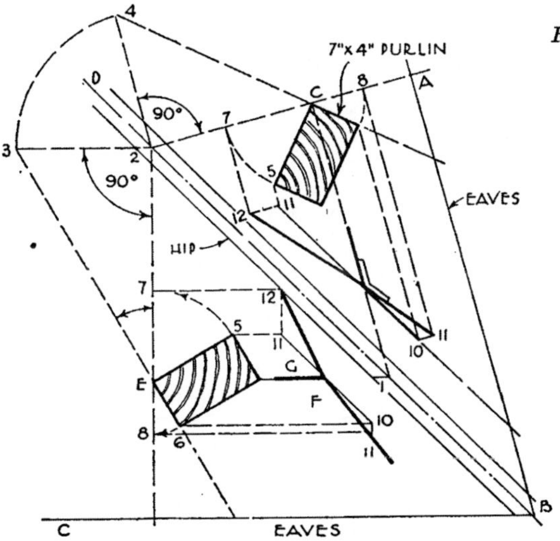

Fig. 9.—Bevels for purlins where hip does not bisect angle, and wall plates are not at right angles to each other

Make 2—4 equal to 2—3; joining C4 should give the pitch of that side of the roof; continue the lines 4C and 3E, and set out the sections and plans as shown in Fig. 4. Then with E as centre and 5 and 6 as radii, obtain the points 7 and 8; projecting down from these points, parallel to CB, points 10 and 11 are obtained, and thus the bevels FG. The bevels for the other side are obtained in a similar manner.

Chapter V

SLATES AND SLATING

Quality, Weight, and Thickness

SLATES are often listed as firsts, seconds, and thirds. This grading refers rather to the thickness and weight than to the actual quality or durability of the slate. A seconds Welsh slate is generally thicker and coarser than a firsts from the same quarry, and hence heavier because it is split from a coarser vein. In practice, the seconds will usually prove quite as durable as the firsts, and perhaps more robust. On this page are a few extracts from the comprehensive Penrhyn slate quarry list, showing weights and prices of first- and second-quality slates of a few of the sizes made by this, the largest of the Welsh quarries. When laid on the roof, the costs of the two qualities work out much the same, the amount saved in initial cost for seconds being counterbalanced by the extra cost of carriage, the seconds being so much heavier per mille than the firsts.

Calculations for Quantities

All slates are laid with a double lap. Each slate overlaps the slate in the next course but one below it (Fig. 4). There are thus two thicknesses of slate in all parts of the roof, and three in most parts.

First Quality				*Second Quality*			
Size	*Computed Weight per 1,200*	*Price per Mille*		*Size*	*Computed Weight per 1,200*	*Price per Mille*	
		Blue	*Green and Wrinkled*			*Blue*	*Green and Wrinkled*
in. in.	*cwt.*	*s. d.*	*s. d.*	*in. in.*	*cwt.*	*s. d.*	*s. d.*
24 × 12	56	700 0	668 0	24 × 12	76	685 0	644 0
20 × 10	36	445 0	420 0	20 × 10	50	462 0	418 0
18 × 9	29	314 0	303 0	18 × 9	40	304 0	298 0
16 × 8	23	247 0	237 0	16 × 8	32	240 0	218 0

Prices are quoted for the purpose of comparison only; they are subject to revision for current rates.

Fig. 1.—OLD DELABOLE SLATING—SLATES OF REGULAR SIZES

Sized Slates

The sizes of slates in common use range from 12 in. by 6 in. up to 24 in. by 12 in. and even 24 in. by 14 in. The practice of specifying any kind of roofing material by its gauge alone is bad, but it is particularly so in slating: it is always better to specify the lap. The standard minimum lap for plain tiling is $2\frac{1}{2}$ in. For slating, roof pitches are on the average lower, and the standard minimum lap has come to be recognised as 3 in. In any specification for slating, the size of slate and the lap should be stated.

The length of a slate on the roof may be considered as made up of twice the gauge plus the lap. In Fig. 4 we have an 18-in. by 9-in. slate laid to a 3-in. lap, and the gauge is therefore (18 in. — 3 in.) $\div$ 2, or $7\frac{1}{2}$ in. To find the number of slates per square we divide the exposed area of each slate (in this case 9 in. by $7\frac{1}{2}$ in.) into one square of 100 ft. super. It is usually convenient in these calculations to work in inches, and our calculation of the number of 18-in. by 9-in. slates laid to a 3-in. lap required to cover a square becomes :—

$$\frac{14{,}400}{7\frac{1}{2} \times 9}, \text{ or } 213.$$

By a similar process it will be found that the number of 16-in. by 12-in. slates laid to a $3\frac{1}{2}$-in. lap required to cover one square is $\frac{14{,}400}{6\frac{1}{4} \times 12}$ or 192.

Calculating the Number of Slates Required

The following Table gives the number of slates of each of the ordinary sizes required to cover one square for $2\frac{1}{2}$-in., 3-in., and $3\frac{1}{2}$-in. laps. It

Fig. 2.—OLD DELABOLE SLATING—RANDOM-WIDTH SLATES OF A STANDARD LENGTH LAID TO A REGULAR GAUGE

also gives the old names for some of the sizes. (These names are rapidly dying out, and nobody really regrets their passing : it is undignified to speak of a Wide Countess or a Narrow Lady. Nowadays no countesses are wide, and all ladies are narrow.) But it is useful to know the names, as they do still appear in some specifications.

Size in Inches	*Name*	*Number per Square When Laid to a*		
		2½-in. Lap	*3-in. Lap*	*3½-in. Lap*
24 × 14	Princesses	96	98	100
24 × 12	Duchesses	112	114	117
22 × 12	Small Duchesses	123	126	130
22 × 11	Marchionesses	134	138	142
20 × 12	Wide Countesses	137	141	145
20 × 10	Countesses	165	169	175
18 × 12		155	160	166
18 × 10	Wide Viscountesses	186	192	199
18 × 9	Viscountesses	206	213	221
16 × 12		178	185	192
16 × 10	Wide Ladies	213	222	230
16 × 9	Broad Ladies	237	246	256
16 × 8	Ladies	267	277	288
14 × 10	Headers	250	262	274
14 × 8	Small Ladies	313	327	343
14 × 7	Narrow Ladies	358	374	392
12 × 6	Small Doubles	505	533	565

Fig. 3.—OLD DELABOLE SLATING—RANDOM SLATES LAID IN DIMINISHING COURSES

How Slates Are Sold—the Cover per 1,000

It is here necessary to digress for a moment to explain the methods by which slates are sold. Sized slates are sold at the quarry by the mille of 1,200. An order to a quarry for 3,500 slates would be interpreted as meaning three and a half times 1,200, or 4,200 actual slates. They are now usually retailed by slate merchants to builders by actual count; that is to say, 1,000 slates are sold as 1,000 actual count. In this article, "mille" means 1,200 and "thousand" or "1,000" means 1,000.

The cover per 1,000 is found by multiplying the exposed area of each slate by 1,000. The cover per 1,000 for a 16-in. by 10-in. slate laid to a 3-in. lap would therefore be :—

$$10 \text{ in.} \times 6\tfrac{1}{2} \text{ in.} \times 1{,}000 = 4{\cdot}52 \text{ squares.}$$

The cover per mille would be :—

$$10 \text{ in.} \times 6\tfrac{1}{2} \text{ in.} \times 1{,}200 = 5{\cdot}42 \text{ squares.}$$

There is a very simple method, not very well known, of finding the cover per mille. Suppose that a 16-in. by 12-in. slate is to be laid to a 3-in. lap. Then the gauge is $6\frac{1}{2}$ in., and the cover per mille is $6\frac{1}{2}$ squares. The cover per mille in squares for a slate 12 in. wide is always the same as the gauge in inches. The reason is easy to see. The cover per mille for a 16-in. by 12-in. slate laid to a 3-in. lap is $12 \times 6\frac{1}{2} \times 1{,}200$ sq. in. =

$$\frac{12 \times 6\frac{1}{2} \times 1{,}200}{14{,}400} \text{ squares.}$$

Cancellation of this fraction leaves only the $6\frac{1}{2}$.

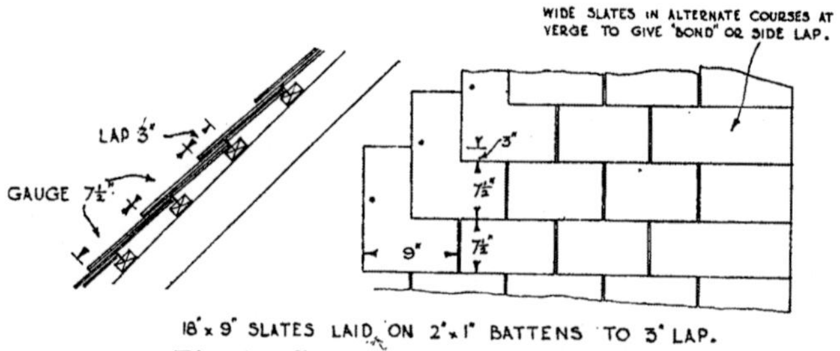

Fig. 4.—GAUGE AND LAP IN SLATING

Then, for a 10-in. slate the cover will be five-sixths of that given by a 12-in. slate of the same length. Hence the cover for a 16-in. by 10-in. slate laid to a 3-in. lap would be five-sixths of 6·5 squares, or 5·42 squares.

One more example will make the point quite clear. What would be the cover in squares per mille of an 18-in. by 9-in. slate laid to a 3½-in. lap ? Gauge = (18 in. − 3½ in.) ÷ 2 = 7¼ in.

Then cover per mille for 18-in. by 12-in. slate = 7·25 squares,

∴ cover for 18-in. by 9-in. slate = ¾ of 7·25, or 5·44 squares.

Random Slates

Random-sized slates are of unequal length and width : generally, the longer a slate the wider it is, so on a roof covered with "randoms" the largest slates, which are always fixed in the lowest courses, may measure 26 in. by 14 in., and the smallest, in the top courses, 12 in. by 6 in. or even smaller. The intermediate courses grade downwards gradually between these extremes.

Slates of random widths have been mentioned above, and the distinction between the two kinds of randoms should be kept clear in the reader's mind, and the points about battening noted. For random-width slating the whole roof can be battened at once : where random-sized slates are to be used the carpenter cannot batten until the slater has given him the spacing for each course, and the slater cannot do this until he has sorted the slates into sizes on the job.

Peggies are small randoms of lengths from 8 in. up to 12 in. or 14 in., and of irregular widths.

Random slates of all kinds are sold by the ton. The covering capacities cannot be worked out by the purchaser, but all quarry owners who make such slates issue lists showing covering capacities per ton. Considerations of space forbid the reproduction of full lists here, but following are short extracts from two such quarry lists. It should be noted that the computed covers per ton are averages, and cannot be guaranteed accurate for any particular consignment.

Examples of Covering Capacity of Westmorland Random Slates

Random sizes, Buttermere Light Green.

Bests : 24 in. to 12 in., in proportionate widths, computed cover 27–28 sq. yds. per ton.

Seconds : 24 in. to 12 in., in proportionate widths, computed cover 21–22 sq. yds. per ton.

Thirds : 20 to 10 in. in proportionate widths, computed cover 15–16 sq. yds. per ton.

The computed *weights* for bests, seconds, and thirds of these slates work out at about 8, 10¼, and 14½ cwts. per square. Note that these slates follow the general rule in slates : the lower the quality, the thicker the slate, so in practice, although the bests given above cost more per ton than the seconds, we get better cover per ton, and the seconds, when laid on the roof, cost about as much as the bests.

The Buttermere Light Green is one of fourteen kinds listed by the Westmorland Green Slate Quarries Ltd., Keswick, Cumberland.

Delabole Randoms

Randoms for laying in graduated courses, Rustic Reds (25 per cent.) and Weathering Grey-Greens (varied mixed shades), 3-in. lap.

No. 1 *Grading :* 24-in. to 12-in. lengths, cover per ton 2·50 squares.
No. 2 *Grading :* 22-in. to 12-in. lengths, cover per ton 2·25 squares.
No. 2*a Grading :* 16-in. to 12-in. lengths, cover per ton 2·25 squares.

Random widths, ordinary grey-green, No. 1 grading :—

Length (× *Random Widths*)	*Cover in Squares per Ton*
20 in.	2·38
18 in.	2·37
16 in.	2·37
14 in.	2·44

Here is a point for careful note. As a general rule, the larger the slate the thicker it is : it is possible to cut a slate of small area to a lesser thickness than one of larger area. This explains why, say, the No. 1 grading in Delabole randoms, ranging from 24 in. to 12 in., gives much the same cover per ton as the No. 2*a* grading, which ranges from 16 in. to 12 in. If all the slates, large and small, were of the same thickness, the No. 1 grading would give by far the better cover, because with large slates less of the roof is covered with three thicknesses of slate than with small slates. The lesser thickness of the small slates just about compensates for the extra area covered with three thicknesses, and as a result the cover per ton given by large and small slates is about equal.

Calculations for Battening

For sized slating and for tiling, the calculations for battening are simple. If battens are fixed on a roof at 12-in. intervals, centre to centre, then on one square of roof there will be 100 ft. run of battens. If the spacing or gauge is 6 in., there will be 200 ft. run to the square. The rule is : Divide 1,200 by the gauge in inches. The following table gives the number of feet run per square for various gauges, and will be found useful for both slating and tiling :—

Gauge in Inches	*Feet run of Battens per Square*	*Gauge in Inches*	*Feet run of Battens per Square*
3½	343	7½	160
3¾	320	8¼	146
4	300	8½	141
4¼	283	9¼	130
4½	267	9½	127
5	240	10¼	117
6	200	10½	114
6¼	192	12	100
6½	185	13	92
7¼	166	13½	89

The battening for random-sized slates cannot be accurately calculated. For 24-in. to 12-in. random slating an average is 190 ft. to 200 ft. per square, but the actual figure depends upon the proportions of large and small slates in a consignment.

Preparation for Slating

The usual methods of preparation for slating are as follows :—

(1) Open battening or rafters. (2) Plain boarding. (3) Plain boarding and battens. (4) Plain boards, felt, and battens. (5) Plain boards, felt, counter-battens, and battens. (6) Untearable felt and battens.

Open battening is more satisfactory as a preparation for slating than for plain tiling. The slates " lie closer " on the roof, and there is little penetration of draught and dirt through a slated roof.

Plain boarding is rather better than open battening, and the appearance from the inside of the roof is improved : this is important where the roof space is used. Plain boarding is hardly ever used in plain tiling, since the nibs of the tiles have to be removed (unless the tiles are nibless), and each tile has to be secured with two nails. In slating, each unit has in any case to be secured with two nails, unless the slates are 12 in. by 6 in. or smaller, when one nail per slate is used.

Methods (3) to (6) are all better than (1) and (2), and one or other should be adopted where cost will permit it.

Holing and Sorting

The nail holes are usually made by the slater, but holing is undertaken by some quarries. The holes are punched on the under side of the slate, at about 1½ in. from each edge. By this method a slight countersinking is formed on the back of the slate—the surface which is uppermost when on the roof—into which the head of the nail fits when the slates are fixed. Without this countersinking the head of the nail would remain above the surface, and the slate above would not " sit down." The position of the nail holes is shown in Fig. 5, from which it will be seen that the position of the holes depends entirely upon the gauge.

Centre nailing, as Fig. 5, is adopted for all except small slates or peggies, which are nailed with one nail at the head. Head-nailing is not safe where the slate is over 12 in. long, as there is risk of lifting in a high wind.

While being holed, the slates should, if necessary, be sorted for thickness. The heaviest slates should be used near the eaves, and the lightest in the top courses.

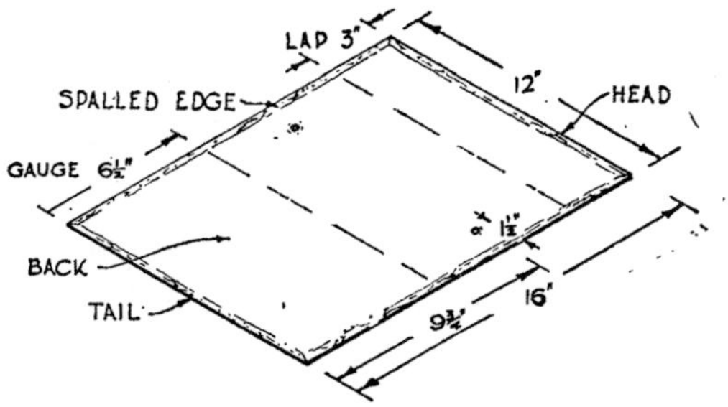

Fig. 5.—Position of nail holes in slate for centre-nailing

Eaves

The eaves course in slating must be formed as in plain tiling, with a double course. Fig. 6 shows the method.

Verges

Ordinary practice in slating is merely to bed and point the verge, but not to fit an undercloak. The use of an undercloak (Fig. 8) is, however, recommended as giving a better appearance from the ground, and a more secure finish. Wide slates should be used at verges and abutments in preference to cutting down an ordinary slate in width. The use of wide slates is, in fact, essential where the roofing slate is 8 in. wide or under (see Fig. 10).

Ridges

The plain angle ridge, or some variation of it, suits a slated roof very well. The angular character of the slate seems to demand an angular finish at ridges and hips. The slates should be brought well up to the ridge tree. Half-round ridge (Fig. 7), either red or blue, may be used, and lead roll is also still popular.

Hips

Angular or other ridge tiles are, in modern practice, widely used on hips. They give excellent cover on both sides, and are not liable to removal in high winds. Lead roll may also be used.

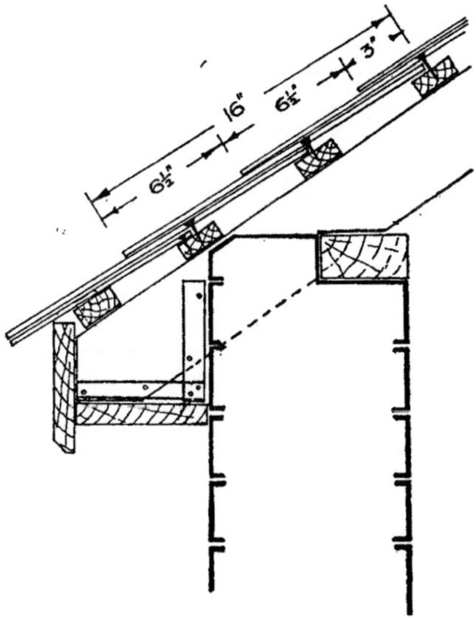

Fig. 6.—Eaves construction in slating

Fig. 7.—Modern slating practice

Showing mitred hip, bedded in coloured cement; half-round ridge; swept valley; vertical slating; slate capping to chimney. The slates are Old Delabole No. 2 grading, 14 in. long, in random widths. (*Setchell & Sons Ltd.*)

In high-class work hips are often mitred, but unless extraordinary care is observed, and a good craftsman employed, the result may be unsatisfactory. The weakness of the mitred hip is that fixing can only be secured along a relatively narrow section of the slate, while the broad base of each mitred slate is exposed. The stripping of a mitred hip in a high wind is a frequent occurrence, particularly where the pitch of the roof is less than 45°, and the building is in an exposed position.

Mitred hips may be either soakered or bedded. The soakers should be of sufficient size to ensure an adequate lap at the head and side, and should be as thin as possible. Thick soakers tend to raise the hip slates, give an unsightly appearance, and assist the wind in its efforts to lift the slates.

Lead soakers are here, as in all other cases, preferable to zinc. It is a good general rule to assume that zinc is satisfactory only where it is completely exposed to the weather. Contact with mortar produces conditions favourable to the disintegration of zinc.

Bedded mitres are preferable, provided that the bedding material is not squeezed out over the face of the slate below. A batten should be fixed on the common rafters along each side of the hip tree to take the edge of the mitred slate. Ordinary glazier's putty may be used for bedding, but there is a number of suitable mastics now available, and coloured cement (Fig. 7) is also used. Too much bedding material is fatal to the appearance of the hip.

Hips may alternatively be flashed, which means, in effect, that the soakers are left exposed instead of being covered by a slate. Seating on turret or similar roofs must be "swept" round, as in Fig. 9. This is best done in small slates.

Valleys

Valleys in slating may be (*a*) mitred, (*b*) open, or (*c*) swept. The mitred valley (Fig. 9) is excellent in appearance and satisfactory in use, provided that the gutter beneath it is properly formed.

Open metal valleys are usual in all but high-class work. It is better to have a fairly wide valley, say 3 in. to 4 in., than a narrow, almost secret, gutter. Secret gutters are usually to be avoided, because of the danger of clogging with dead leaves. The wide valley ensures always a clear passage for water. Open valleys may be "left dry," or an undercloak may be fitted: the latter method is preferable. An undercloak 4½ in. wide is ample.

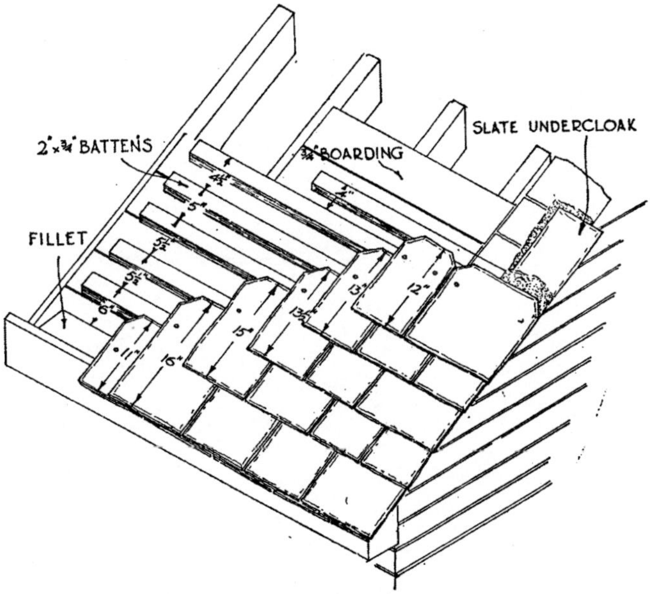

Fig. 8.—BOARDING, BATTENING, AND RANDOM SLATING; SLATE UNDERCLOAK AND WIDE SLATES IN ALTERNATE COURSES AT VERGE

Swept valleys in slating (Fig. 7)

are unusual and expensive. They can be gracefully formed only in slates of narrow width, as wide slates come round the valley in, as it were, a series of jerks, with wide spaces between the slates. With small slates a smooth curve can be formed. Large slates, again, require that the valley shall have a very wide sweep; with small slates the curve may be more restricted. The wide curve necessary with the large slate creates a difficulty at the top of the valley, and the junction with the ridge is likely to be ungainly.

Fig. 9.—12-IN. BY 6-IN. OLD DELABOLE SLATES ON A CIRCULAR TURRET ROOF; MITRED VALLEY

Abutments

Soakers must always be used at abutments, and the width along the roof must always be sufficient to give the proper side lap. The turn-up against the wall must also be enough to ensure adequate cover by the stepped or other flashing.

Fig. 10.—USE OF WIDE SLATES AT VERGES OR ABUTMENTS

Nailing

The question of nails for both slating and tiling is discussed in the next chapter. Here we need only remind the reader that the life of a slated roof is only as long as the life of the nails. Galvanised nails should not, therefore, be used on any slated roof.

Roof Pitches

The pitch of the rafters may generally be less for a slated than for a plain-tiled roof, but the size of slate must be taken into account. Large slates may be laid on a slower pitch than small slates, and the reasons are worth remembering. There is always a certain spread of moisture on either side of the vertical joints between adjacent slates in the same course, and the slower the pitch the greater the spread. The side lap in slating is equal to half the width of the slate, and if a small slate is used on a roof of slow pitch the moisture may spread until it finds its way between the vertical joints that are covered, and so into the roof.

A further reason for using large slates only on slow-pitched roofs is that the weight to be borne by the nails is reduced to a minimum. A 24-in. by 12-in. slate may weigh up to 7 lb., and this is too great a weight to be borne by two nails on a slope of 50° or 60°

Random Slating

Fig. 8 is an illustration of random slating on boards and battens. Nearly all random slates are shouldered at the quarries, on account of the difficulty of ensuring trueness to width in this class of slate. The slates shown diminish from 16 in. to 12 in., and the gradual decrease in the gauge to give a 3-in. lap throughout should be noted.

Chapter VI

HOW TO TILE ROOFS

Clay Tiles

CLAY suitable for the manufacture of roofing tiles is widely distributed over England, the largest deposits being in the Broseley district of Shropshire, North and South Staffordshire, Berkshire, the Bridgwater district of Somerset, and the south-east corner of England. The clays vary in their mineralogical and chemical compositions, and it is these variations that account for the different characteristics of the tiles produced. The Midland clays, those of Broseley and Staffordshire, for example, contain a high percentage of iron, and the tiles made in those districts can be naturally burnt to purple, brown, and blue colours that cannot be approached in tiles made in some other parts of the country, except by the application of artificial colouring.

Again, some clays are eminently suitable for the manufacture of only plain tiles, and others can be used as well for the production of larger units such as pantiles, Roman tiles, and single-lap tiles generally. It is, in fact, considered that tiles such as the Bridgwater Double Roman, Marseilles, and other types that depend for their effectiveness upon the roof on trueness to shape, can be made only from alluvial clays of recent deposit. However that may be, it is true that such tiles are successfully made only from those clays.

A discussion of the technicalities of clay, and the suitability of each kind for the manufacture of either bricks or tiles or both, is outside the scope of this book, but the reader should remember that there are nearly always very good reasons for the production of one kind of tile instead of another in any particular district. These reasons are usually technical, but often they form part of the fascinating history of the development of the English roofing tile.

Qualities

When taken from the kiln, clay tiles are sorted into best, second, and third qualities. Best tiles should be without a fault of any kind; seconds tiles, as sent out by most works, are those which have some fault which precludes their ranking as best tiles, but which is not detrimental to the tile as a roof covering. Tiles that are slightly out of true, or have an unsightly " flash " or other blemish on their surface, are classed as

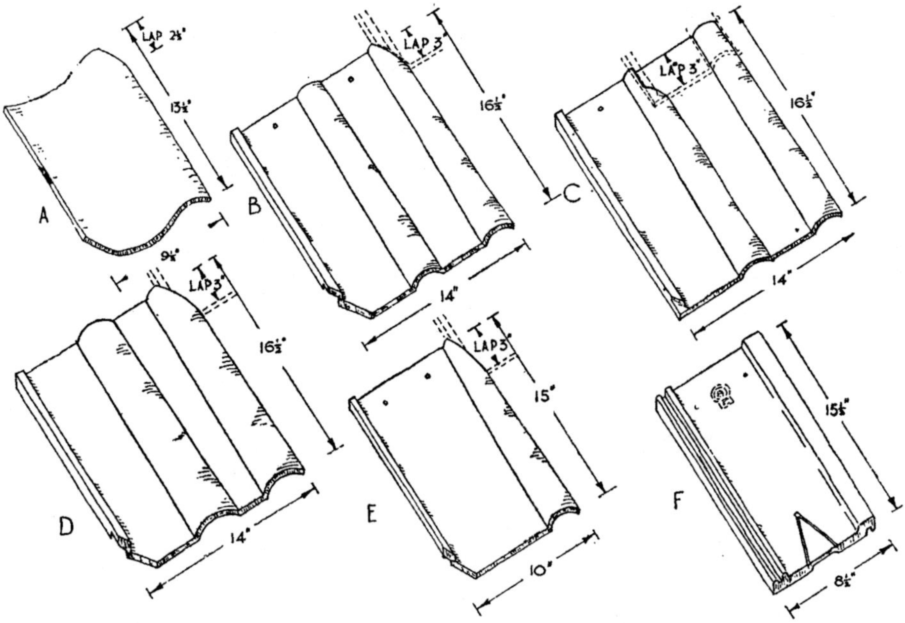

Fig. 1.—A FEW POPULAR FORMS OF ENGLISH SINGLE-LAP TILE

A. Old English pantile. B. Double Roman tile—straight joint. C. Double Roman tile—broken joint. D. Reynardo tile. E. Single Roman tile. F. Somerset interlocking tile—No. 13.

seconds. Third-quality tiles are usually either badly twisted or fire-cracked " above the gauge line," as the tiler says.

Classification of Clay Tiles

For practical purposes, tiles are most conveniently classified as single-lap and double-lap or plain.

In single-lap tiling each tile overlaps or is overlapped by the tiles on either side of it in the same course, and the tiles in the courses above and below. There is thus only one thickness of tile in most parts of the roof, and two thicknesses in some parts. Single-lap tiles may have a simple overlap, as in the case of the English pantile (Fig. 1), or there may be an interlock at the head or at the side and head. The Somerset interlocking No. 13 (Fig. 1) has a side lock. The imported interlocking improved pantile (Fig. 4) has both side and head locks.

In double-lap or plain tiling, each tile overlaps the tile in the next course but one below it, but there is no overlap at the side. The length of tile exposed on the roof is called the gauge, and the amount by which one tile overlaps the tile in the course next but one below it is called the lap. In plain tiling, it will be noted, there are at least two thicknesses of tile in every part of the roof, and three thicknesses in most parts.

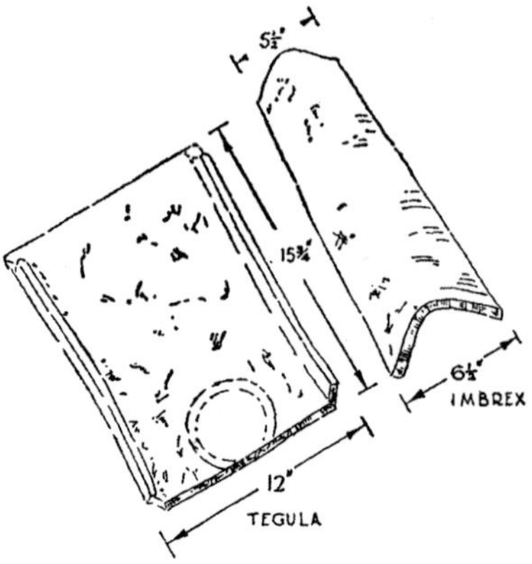

Fig. 2.—ROMAN TEGULA AND IMBREX

Illustrations of a number of kinds of single-lap tiles are given in Figs. 1 to 5, and in the notes that follow a brief description of each will be found. The approximate sizes, covering capacities, and gauge for battens of each are given. Users are advised to wait until tiles are delivered before fixing the battens, as in certain cases slight variations in length as between one consignment and another may occur.

Old English Pantile

The English pantile (Fig. 1) is a descendant of the Spanish tile (see Fig. 3), also called the over-and-under tile and Spanish Mission tile. It is probable that the Flemings first evolved this single tile, consisting of a channel and a roll, out of the two units that make up the Spanish tile. The pantile is justly popular, as it fits admirably into the modern architectural picture. It is made in many parts of the country, Bridgwater, Berkshire, and East Anglia being the largest centres of production.

The following data are subject to much variation, as the pantiles made by different works are dissimilar in both length and width.

Size: 13½ in. by 9 in. Cover: 150 per square. Gauge for battens: 12 in.

Double Roman Tile

The Double Roman tile (Fig. 1) is a product of the Bridgwater district of Somerset, and is the largest tile in use in England. It is a descendant of the Roman *tegula* and *imbrex* (Fig. 2), which the Romans used extensively in England for their villa roofs. In the Double Roman tile, two *tegulæ* and two *imbrices* are combined into one unit.

Technical Data.—Size: 16½ in. by 14 in. Cover: 85–90 per square. Gauge for battens: 13½ in.

Single Roman Tile

This is the smaller, and not so popular, brother of the Double Roman tile, and is a little closer to the line of descent from the *tegula* and *imbrex*. It also is a product of Bridgwater, although a nearly similar type is made in Lincolnshire. A larger type, equal in size to the Double Roman, is also made.

Technical Data.—Small Pattern. Size: 15 in. by 10 in. Cover: 125 per square. Gauge for battens: 12 in. *Large Pattern.* Data as for Double Roman.

Double Roman Interlocking

This tile has the same appearance as the ordinary Double Roman tile on the roof, but is laid with a broken joint. The ordinary Double Roman is laid with a straight joint. (See Fig. 1.) The practical advantage of the broken joint is that the water running down the side channel of any one tile is discharged on to the face of the tile below, and not, as in the case of the straight-joint tile, into the side channel of the tile below.

Technical Data as for Double Roman tile.

Somerset Interlocking Tile No. 13

A very effective tile (Fig. 1), manufactured expressly for cheap domestic or industrial roofing. It is of adjustable lap, and has been used without felt on roofs as low as 30° of pitch at the eaves. This tile is produced by the Somerset Trading Co. Ltd., Bridgwater.

Technical Data.—Size : 15½ in. by 8 in. Cover for 3-in. lap, 155 tiles per square. Gauge for battens for 3-in. lap : 12½ in.

Reynardo

The Reynardo (Fig. 1) is essentially a Double Roman tile, but it has a bolder roll, and on the roof an appearance somewhat similar to that of the Spanish tile. The Reynardo tile is an exclusive product of Messrs. Colthurst, Symons & Co. Ltd., Bridgwater.

Technical Data as for Double Roman tile.

Spanish Tile

The tiles are laid alternately concave and convex in courses, and there is a lap at both sides and at the head of each tile (Fig. 3).

Technical Data.—There are many makes of Spanish and Italian tile, all of different size and covering capacity, and users should obtain information from the manufacturers concerned.

Italian Tiles

Similar in type to the Spanish tile, with this difference—that the under tile is flat instead of concave. This distinction between Italian and Spanish tiles is not generally understood, and cases have occurred where a specification has called for Spanish tiles when Italian tiles were meant.

Both kinds are expensive, but eminently suited to buildings where a bold, vertical roll on the roof is required to heighten the effect of the general elevation.

Courtrai-du-Nord Tile

This tile, shown in Fig. 5B, was once known in England only as a cheap but effective roof covering. It is now made in a great variety of colours, both matt and glazed : meadow green glazed, bluish green glazed, tea-

Fig. 3.—SPANISH TILING—SHOWING USE OF TAPERED HIPS. (*Colthurst, Symons & Co. Ltd.*)

pot brown glazed, autumn leaf glazed, moss green, and lichen green are a few of the colours listed by the importers, Messrs. Langley London Ltd.

Technical Data.—Size : $11\frac{1}{2}$ in. by $8\frac{1}{2}$ in. Cover : approximately 208 tiles per square. Gauge for battens : approximately $9\frac{3}{8}$ in.

Improved Interlocking Pantile

This (Fig. 4), like the Courtrai-du-Nord tile, is of Continental make, and obtainable in a variety of colours, both matt and glazed. Comparison with Fig. 2 shows that it differs from the ordinary English pantile principally in the fact that the interlocking improved pantile has a side and head lock. For technical reasons this type of pantile has not so far been made in England.

Technical Data.—Size : $13\frac{1}{4}$ in. by $9\frac{3}{4}$ in. Cover : approximately 147 tiles per square. Gauge for battens : approximately $11\frac{1}{2}$ in.

The Cloister Tile

This is designed to give on the roof an effect somewhat similar to that of Spanish or Italian tiling (Fig. 5A).

Technical Data.—Size : $11\frac{3}{4}$ in. by $8\frac{1}{2}$ in. Cover : approximately 200 tiles per square. Gauge for battens : approximately $9\frac{7}{8}$ in.

Plain Tiles

Fig. 6 illustrates the ordinary two-holed, two-nibbed plain tile, and the fittings that are used with it for roofing and vertical tiling. Two sections of half-round ridge are shown : 90° pitch for main ridges, and 105° for hips. The 90° ridge is too bold for use on hips, and its use for that purpose makes an awkward intersection at the point where the main ridge and hip meet.

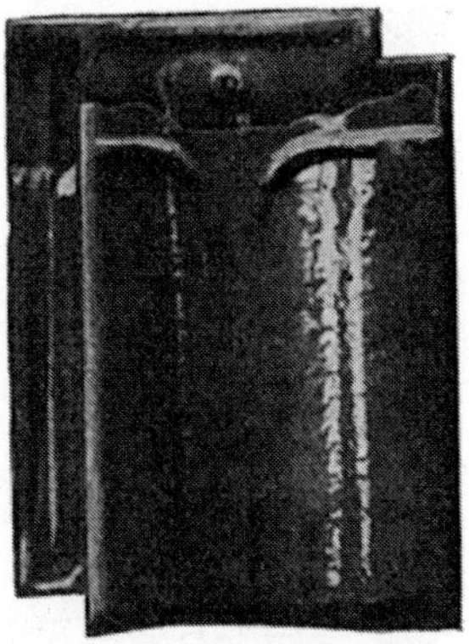

Fig. 4.—IMPROVED INTERLOCKING PANTILE

Fig. 5A.—CLOISTER TILE

Fig. 5B.—COURTRAI-DU-NORD TILE
(*Langley London Ltd.*)

CALCULATIONS

Plain Tiling

It is commonly said that a square of tiling requires 600 plain tiles, and although this is a good approximate figure, it is rarely accurate. The standard size of plain tiles is 10½ in. by 6½ in., but in practice sizes vary from 10 in. by 6 in. up to 11 in. by 7 in., and to calculate accurately the number of tiles needed to cover one square we must know (*a*) the size of tile, and (*b*) the lap, or gauge.

The length of each tile when laid on the roof may be considered as made up of twice the gauge plus the lap, and a 10½-in. by 6½-in. tile laid to a 2½-in. lap shows a gauge of 4 in. The number of tiles required to cover one square is clearly the number of times that the exposed area of each tile divides into one square. Thus :—

$$\frac{1 \text{ square}}{6\tfrac{1}{2} \text{ in.} \times 4 \text{ in.}} = \frac{14{,}400 \text{ sq. in.}}{6\tfrac{1}{2} \text{ in.} \times 4 \text{ in.}}$$

$$= 554 \text{ tiles.}$$

Similarly, an 11-in. by 7-in. tile laid to a 3½-in. lap (3¾-in. gauge) will require :—

$$\frac{14{,}400}{7 \text{ in.} \times 3\tfrac{3}{4} \text{ in.}} \text{ or } 549 \text{ tiles.}$$

An 11-in. by 7-in. tile laid to a 2-in. lap (4½-in. gauge), as for vertical tiling, will require only 457 tiles per square.

The following Table gives the number of plain tiles per square required for three sizes of tile laid to various laps. A minimum lap of 2½ in. is recommended for roofing, but 1½ in. is usually sufficient for vertical tiling.

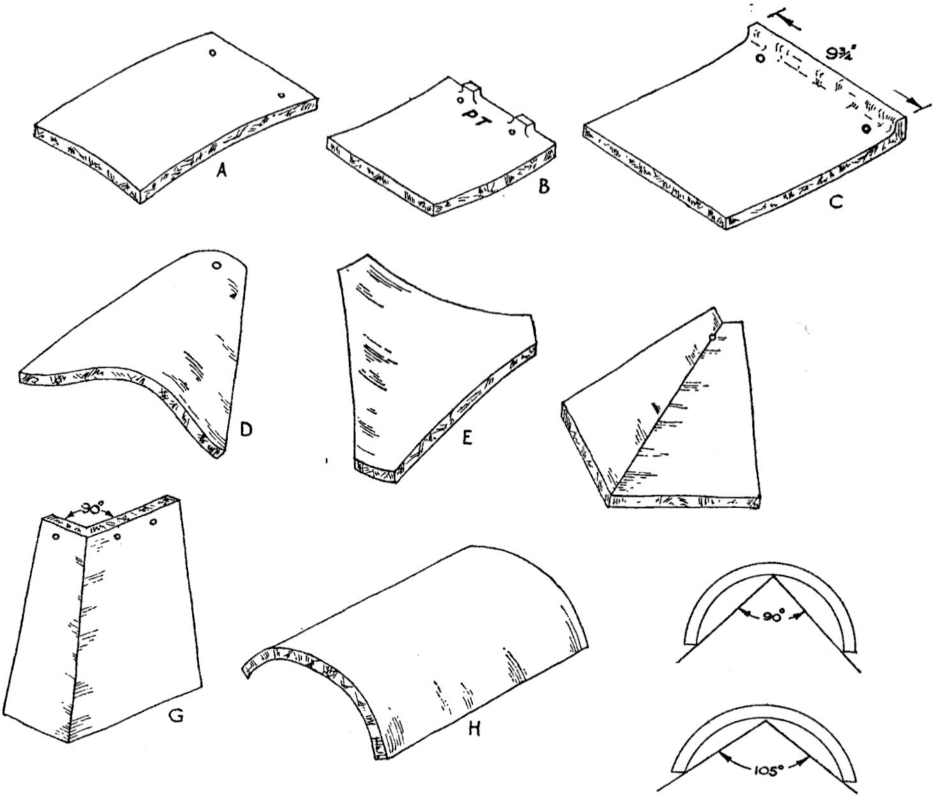

Fig. 6.—Plain tile and fittings

A. Plain tile. B. Under side of eave tile. C. Under side of tile-and-half tile (continuous nibbed, machine made). D. Bonnet hip tile. E. Valley tile. F. Close-fitting hip tile. G. Angle tile handed, external. H. Half-round ridge tile.

Size of Tile	*Lap in Inches and No. per Square*					
	1½	2	2½	3	3½	4
11 in. by 7 in.	433	457	484	514	549	588
10½ in. by 6½ in.	492	521	554	591	633	682
10 in. by 6 in.	565	600	640	686	738	800

Ridge Tiles

These are usually made in 12-in. or 18-in. lengths, although odd lengths do sometimes occur. The required number is easily obtained.

Bonnet Hips, Hip Tiles, and Valley Tiles

A rough and ready method of calculating the number of these is to take 3 per foot run of hip or valley, but this figure in many cases will allow rather too many for waste. In Fig. 7 the length of the common rafter is 16 ft., and of the hip, 20 ft. The tiles are laid to a 4-in. gauge. If we allow 3 hip tiles per foot, the number ordered would be 60, or 240 for the four hips. The actual net number required for each hip would be 48 or perhaps 49. This is estimated as follows: the number of hip tiles will be the same as the number of courses of tiles on the main slope: a rafter length of 16 ft. gives us, at 4-in. gauge, 48 courses of tiling, and this is the number of hip tiles that will be needed. For the four hips the number would be 49 by 4, or 196, and the tiler would order, say, 204.

The calculations for valley tiles are made similarly.

Angle Tiles

The number of these is estimated similarly by finding the number of courses of plain tiles.

Single-lap Tiling

Here the principle is the same as for plain tiling: divide 100 sq. ft. by the exposed area of each tile, but it must be remembered that the length of each single-lap tile is made up of the gauge plus the lap.

Example.—A single-lap tile measures 16½ in. by 14 in., and has a vertical lap of 3 in., and a side lap of 1½ in. The number of tiles required per square is :—

$$\frac{14{,}400}{13\frac{1}{2} \times 12\frac{1}{2}} \text{ or 85 tiles.}$$

These figures are, in fact, those applicable to one of the best known of the English single-lap tiles: the Bridgwater Double Roman tile. The technical data given in the preceding pages for each kind of single-lap tile named have been similarly calculated. Makers or importers will always give this information to customers: the figures are, in fact, almost invariably given in their catalogues.

Weights

The weights of plain tiles are approximately as under :—

10 in. by 6 in.	19–21 cwts. per 1,000
10½ in. by 6½ in.	20–23 cwts. ,, ,,
11 in. by 7 in.	21–25 cwts. ,, ,,

but it is advisable always to get the correct weight from the makers.

Bonnet hips weigh about 3½ cwts. per 1,000; angle tiles about 4 cwts. per 1,000; and ridge tiles, 12 in. long, about 8½ lbs. each.

CHOOSING A TILE

In choosing a tile, the architect has to decide (*a*) the type of tile, whether plain or single-lap, and sometimes (*b*) the kind of tile, whether hand- or machine-made. The following hints may be of assistance to those who have to decide how the roof, the most important single element in an elevation, is to be covered.

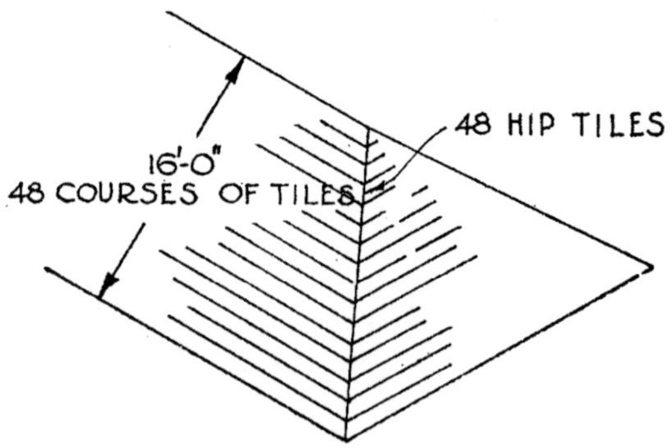

Fig. 7.—ILLUSTRATING METHOD OF CALCULATING NUMBER OF HIP TILES

(1) In single-lap tiling the unit is large, and usually shows a series of well-defined vertical and horizontal lines on a roof. Single-lap tiles are therefore suitable for large and high buildings.

(2) The plain tile is suitable for complicated roofs, having hips, valleys, dormers, and small projections.

(3) Concrete tiles are usually cheaper than machine-made clay tiles of comparable quality, and the latter are cheaper than hand-made tiles.

(4) For slow-pitched roofs the single-lap tile is to be preferred.

(5) Care should be taken to ensure that the tile used is one that can at any time be repeated. Do not, for example, use an imported single-lap tile of which future supplies are not assured.

(6) Remember that if a hand-made tile is chosen, it will " weather " on the roof, but not on the vertical tiling. Machine-made tiles, unless sand-faced, do not usually weather.

(7) If a machine-made clay tile is chosen, one that has been burnt naturally to a dark or brindled shade is preferable to a light-red tile. The latter may laminate ; the former will not. It is not, however, possible to be dogmatic on this point, for the products of some works are light red, however long the period of burning may be.

(8) A well-cambered plain tile should be chosen in preference to a tile that is straight in its length (Fig. 8). The camber not only helps the tail of each tile to sit snugly on the course below, but it provides an air space for ventilation among the tiles and battens. Lamination is hastened by the close bedding of one tile on another for the greater part of its length : capillary action tends to keep the upper surface of one and the lower surface of the other damp for long periods. The circulation of air among plain tiles is assisted if the tile is slightly cambered in its width. Fig. 8 shows a plain tile cambered only in its length, and two kinds of tiles cambered in their length and width.

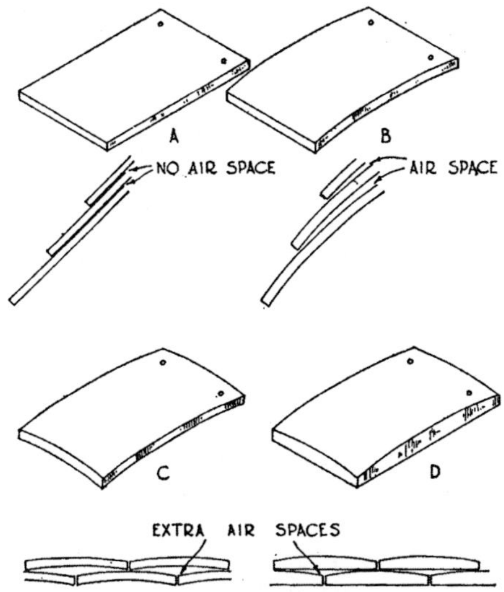

Fig. 8.—A WELL-CAMBERED PLAIN TILE SHOULD BE CHOSEN IN PREFERENCE TO A TILE STRAIGHT IN ITS LENGTH

A. Plain tile straight in its length and width. B. Plain tile cambered in its length only. C. "Full" double-cambered tile. D. "Half" double-cambered tile.

Special Tiles

It often happens that certain fittings—bonnet hips or angle tiles, for example—have to be made to a special pitch for a particular job. In such cases it is advisable to give the works as long notice as possible. Special fittings cannot usually be produced in less than eight weeks from the date of order, and contracts are frequently delayed because the order for the tiles is placed shortly before the roof is ready. It may even happen that, particularly in single-lap tiling, the tiles for part of the roof have to be purposely made. Fig. 9 shows a roof, circular on plan, to be covered with hand-made pantiles: on each slope there have to be the same number of tiles at the eave as at the ridge. Hence, on the outer slope the pantiles have to be wider at the tail than at the head, and vice versa on the inner slope. Special moulds have had to be made for each course—sixteen on each slope.

Concrete Tiles

The concrete tile has taken its place beside the other and longer-established roofing materials. The quality of the best-known makes is high, the permeability is negligible, and the colours, although not yet permanent, do not bleach out and disappear entirely.

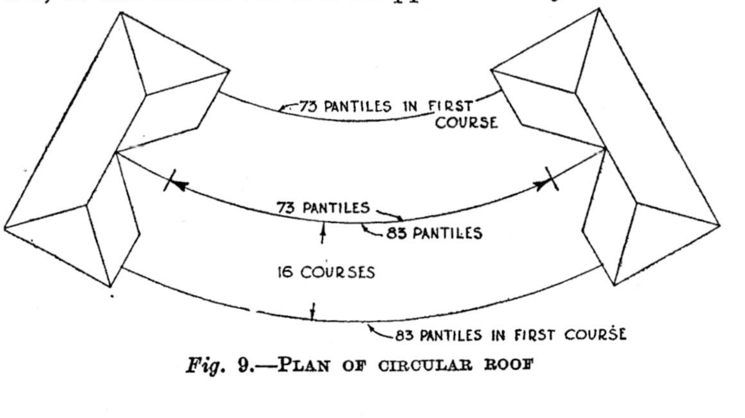

Fig. 9.—PLAN OF CIRCULAR ROOF

Most concrete tiles are of the plain-tile pattern, but some firms make a form of interlocking tile that is, in appearance on the roof, similar to the Somerset No. 13 tile, while one firm, the Marley Tile Company, manufactures concrete single-lap tiles of the Double Roman, Reynardo, pantile, and Spanish patterns.

The methods of application of concrete tiles are the same as those described for clay tiling : the concrete tile has, in fact, been made directly in imitation of the clay tile. The usual fittings—ridges, hips, valleys, angle tiles—are all of the same patterns as those used with clay tiles.

TILING WITH PLAIN TILES

One striking difference between old and new methods should be noted at the outset. Until recent years it was the custom to try to prevent the penetration of water by the lavish use of mortar in the form of torching and bedding. In modern practice the use of mortar is deprecated. We ventilate, rather than seal, our roofs, and there can be no doubt that it is the better practice. The life of the tiles and battens is prolonged, and the drying out of wet tiles is accelerated, by a current of air passing between them.

Roof Pitches

The question of rafter pitch for plain tiling must always be considered in conjunction with the other factors involved : permissible cost, position and exposure of building, and kind of tile.

The cost of all items increases as the pitch is raised, and this is a factor to which architects have to give consideration. A sheltered roof can safely have a slower pitch than could be used for one in a very exposed position. In general, plain tiling demands a steeper pitch than does any form of single-lap tiling, but that is not to say that the appearance of a house with a single-lap tiled roof will not be improved by having a roof of fairly steep pitch rather than one of 30°.

Experience has shown that in England a pitch of 45°, there or thereabouts, is satisfactory for a plain-tiled roof under ordinary conditions of exposure. Greater security can be obtained in exposed positions by (*a*) increasing the lap or (*b*) raising the pitch.

A point to be noted is that in plain tiling the effective pitch (Fig. 10) is several degrees less than the pitch of the rafters, particularly when, as in the illustration, a well-cambered tile is used. It is of the greatest importance that this difference should be remembered in connection with valleys. If the pitch of a common rafter is 40°, the pitch of the valley rafter is 31°, and a well-cambered valley tile used on such a roof may have an effective pitch of as little as 20°. With a swept or laced valley the risk of penetration is even greater than with a valley tile, and a generous width of felt is advisable with swept or laced valleys if the common rafter has a pitch less than 45°, particularly if the tile is a relatively porous, hand-made one.

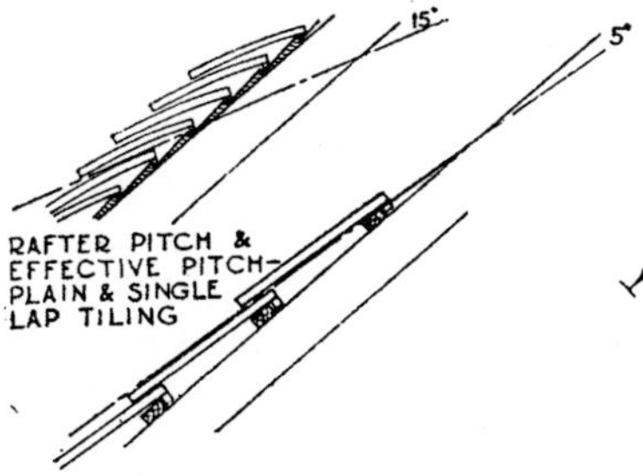

Fig. 10.—RAFTER PITCH AND EFFECTIVE PITCH

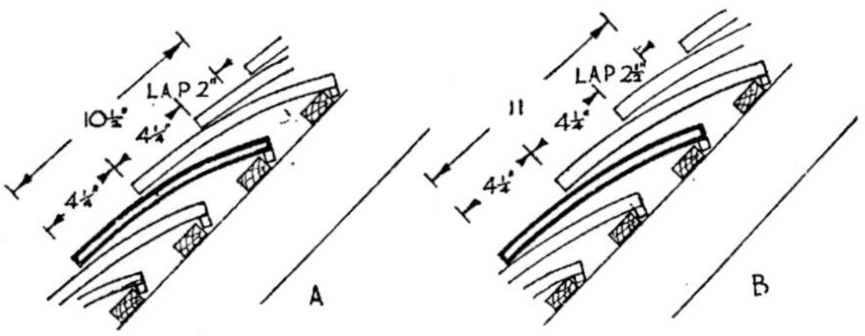

Fig. 11.—GAUGE AND LAP—11-IN. AND 10½-IN. PLAIN TILES

Gauge and Lap

The lap should always be mentioned in any specification, since tiles vary in length, and to specify only the gauge is to run the risk of getting insufficient lap. Two and a half inches is the minimum lap for plain tiling, and, as Fig. 11 shows, while this lap is obtained with a 4¼-in. gauge where a tile 11 in. long is used, a 10½-in. tile would give only a 2-in. lap.

Roof Preparation

The following are the alternative forms of roof preparation adopted in modern practice :—

(*a*) Open battens : ideal from the point of view of roof ventilation, but obviously unsuitable where use is to be made of the roof space.

(*b*) Feather-edged boards : less ventilation and less dust and soot in the roof space than with (*a*), and rather better " tying-in " value ; should never be used with a relatively porous, hand-made plain tile.

(*c*) Plain boards and battens : better than either (*a*) or (*b*), and giving, with good boarding, a practically dustproof roof space. The addition of counter-battens is well worth the small extra cost.

(*d*) Untearable felt and battens : comparable in value with (*c*), but the felt must be untearable (that is, a bituminous felt with a strong canvas or other base), or the tilers will put their feet through it.

(*e*) Plain boards, felt, counter-battens, and battens : the ideal roof preparation from all points of view, but the most costly.

Sizes of Battens

For plain tiling, battens may be of any scantling from 1 in. by ¾ in. up to 2 in. by 1 in. A good ordinary size is 1¼ in. or 1½ in. by 1 in. In plain tiling, the gauge should be taken from top to top of battens, not from centre to centre.

Eaves

A section through the eaves in a plain-tiled roof is shown in Fig. 12A. This also illustrates a normal method of closing the soffit of overhanging

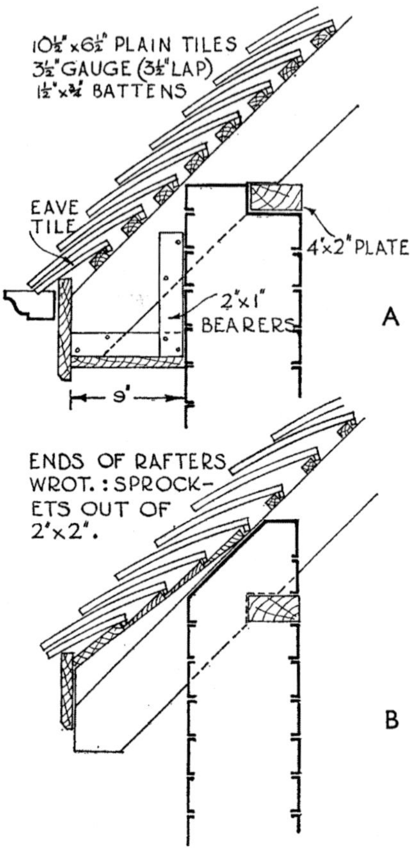

Fig. 12.—Sections through eaves in plain tiling

rafters. Fig. 12B shows how an open soffit should be formed. Note the use of feather-edged boarding in place of battening from eave to wall-line. Eaves are occasionally specified as "to be bedded," but this is unnecessary and detrimental. In Fig. 13 the construction at eaves is clearly shown, the batten for the eave tile being the first above the fascia. The tiler's method of loading the roof is also seen: a tiler's mate will carry up a ladder as many as forty to forty-five plain tiles on his head, a weight of perhaps 100 lbs. or more.

Sprocketed eaves improve the appearance of a roof, but care should be taken that the sprocketed part is not of too slow a pitch. No part of a plain-tiled roof, excepting, of course, hips and valleys, should be of less than 40° pitch.

Nails and Nailing

Nails for tiling and slating may be conveniently considered under one heading.

The kinds ordinarily in use are of galvanised iron or steel, zinc, yellow cast composition, and copper.

Galvanised nails are used for ordinary tiling work, where the weight of the unit is largely taken by the nibs. For slating or nibless tiling, where the nail takes the whole weight, galvanised nails are considered to be unsuitable. Although gauges as fine as 12 and 13 are frequently used, a minimum of 11 gauge is recommended. The length should be at least 1½ in. for tiling: a nail 1¼ in. long, if used with a thick tile, will not give sufficient fixing into the battens.

Zinc nails are cut, not cast, and are satisfactory on all jobs excepting those where the nails may be exposed to fumes that are injurious to zinc. A minimum length of 1¼ in. for small slates, and 1½ in. or 2 in. for large slates, is recommended. Average weights of zinc nails are :—

1½ in.	4½–5 lbs. per 1,000
1¾ in.	6½–7 lbs. ,, ,,
2 in.	10–11 lbs. ,, ,,

Fig. 13.—Construction at eaves in plain tiling

Yellow cast-composition nails are made of an alloy of copper, zinc, and tin. They are practically indestructible under all conditions, and have the advantage over copper that they are harder and more readily driven. Composition nails are used for all kinds of slating and tiling, but are costly: about 2*d*. per pound more than copper nails, and about four times as much as galvanised. Composition nails are about 10 per cent. heavier than copper.

Copper nails are generally used for slating, and for this purpose are unequalled. Of the many slated roofs that have to be stripped and re-slated, in probably 95 per cent. of the cases the trouble is due to defective nails rather than to any defect in the slates themselves. Galvanised nails should never be used in slating: copper, zinc, or composition are safer. Average weights of cut copper nails are :—

$1\frac{1}{2}$ in. . .	$4\frac{1}{2}$ lbs. per 1,000
2 in. . .	10 lbs. ,, ,,

Cast copper nails are lighter than the cut variety.

Cost of Nailing

It is not possible to give costs per square for nailing: each class of work requires a different number of nails per square. Plain tiling is ordinarily nailed at every fourth or fifth course, with two nails to each tile. Fifth-course nailing laid to a 4-in. gauge thus requires 554 × 2 ÷ 5,

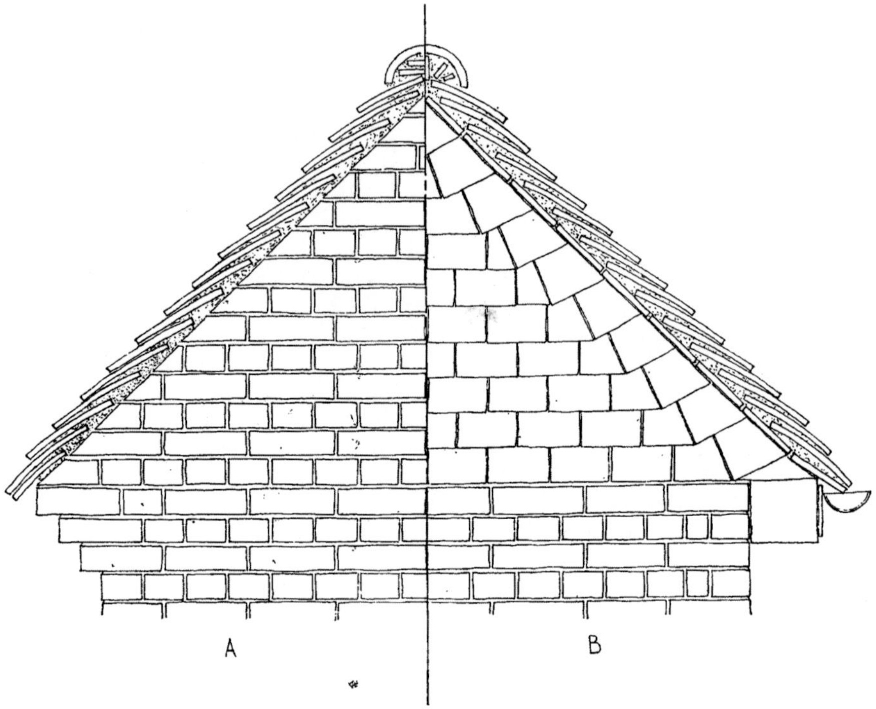

Fig. 14.—Verges in plain tiling

A. Verge bedded and pointed. B. Bedded and pointed, with course of plain tiles a undercloak. Note Winchester cut of vertical tiling against undercloak.

or 222 nails per square net, and if the nail used is a copper nail $1\frac{1}{2}$ in. long, weighing $4\frac{1}{2}$ lbs. per thousand, and costing 10*d.* per pound, the cost per square for nails would only be about 11*d.* This would allow a reasonable percentage for waste. The 554 in the above calculation is the number of tiles per square. Other examples are worked out by the same method.

Verges

Verges ordinarily project 2 in. over a brick wall or barge board, but on high buildings this should be increased. Verges may be bedded and pointed only (Fig. 14A), or bedded and pointed with the addition of one or more courses of plain tiles laid flat as undercloak (Fig. 14B). The latter method is preferable for two reasons : the appearance is improved, and the verge is given a slight tilt, which helps to prevent water from dripping over the edge of the verge.

Where vertical tiling occurs under a verge, as in Fig. 14B, the undercloak is essential if a neat finish is to be made between the vertical tiling and

undercloak. This illustration also shows the method of splay-cutting known as Winchester cut, a finish that is becoming increasingly popular. It is, of course, more expensive than the continuation of the courses of vertical tiling in a straight line until they meet the under side of the verge.

Figs. 15 *and* 16.—HIP COVERED WITH HALF-ROUND RIDGE

Note use of lead where hip passes near chimney stack.

Ridges

Ornamental ridges are now rarely used in tiling: nor, for that matter, in slating either. Most architects favour, for tiling, the half-round ridge, bedded and pointed in cement mortar. The practice of bedding ridges, hips, etc., in lime and hair mortar, and pointing in cement mortar, is also dying out. It had no practical advantage, and was rarely carried out to specification.

Fig. 17.—Bonnet hip tiling

Hips

Hips may be :—

(*a*) Covered with half-round or other ridge tiles (Figs. 15 and 16).

(*b*) Of bonnet hips (Fig. 17).

(*c*) Of hip tiles (Fig. 18).

(*d*) Mitred (Fig. 19).

Half-round ridge may be used with either hand-made or machine-made tiles. Bonnet hips are usual in good-class hand-made tile work, and hip tiles are associated with machine-made tiles. Mitred hips are commoner in slating than in tiling.

Here are a few points to be noted :—

(1) Hip irons are needed with a ridge on the hip. Their function is to prevent the whole length of hip ridge from slipping.

Fig. 18.—Hip tiling with close-fitting hip tiles

(2) Every bonnet hip or hip tile has to be nailed, and there is thus no necessity for hip irons.

(3) There is one bonnet hip or hip tile to each course of tiles on either side of the hip. If the two pitches coming into a hip are different, the longest common rafter on the steeper pitch will be shorter than that on the slower pitch, but the number of

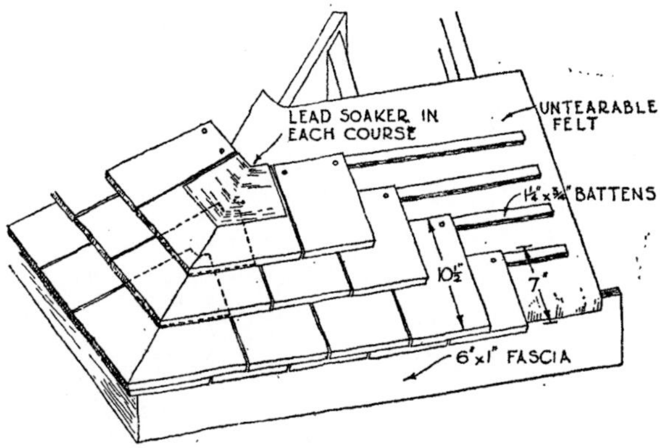

Fig. 19.—Mitred hip in plain tiling

courses of tiling remains the same. Therefore, on the steeper pitch the gauge is shorter, and provision must be made for this when battening. The rule is: where bonnet hips or hip tiles are used, the battens must intersect at the hip.

These remarks apply also to valley tiling and mitred valleys and hips. Where ridge tiles are used on a hip, or valleys are of metal, the courses need not "follow round."

Valleys

The valley is the most vulnerable part of a roof, and needs extra care on the part of the tiler. The usual forms of valley are (*a*) valley tile, (*b*) mitred valley, (*c*) laced valley, (*d*) circle or swept valley, and (*e*) metal valley.

The valley tile (Fig. 21) is the commonest and most serviceable form for use with plain tiling. Valley tiles are not nailed, as the tiles on either side keep the valley tile in place.

The mitred valley is not much used in modern practice, and the open valley only in cheap work. Both involve the co-operation of another trade, and both are less satisfactory than the tiled valley.

Laced and swept valleys are pleasing in appearance, but expensive in both labour and material, as so many tiles have to be cut. A laced valley is shown in Fig. 22. Note the characteristic upward sweep of the courses as they approach the valley, and compare with the tiled valley of Fig. 21.

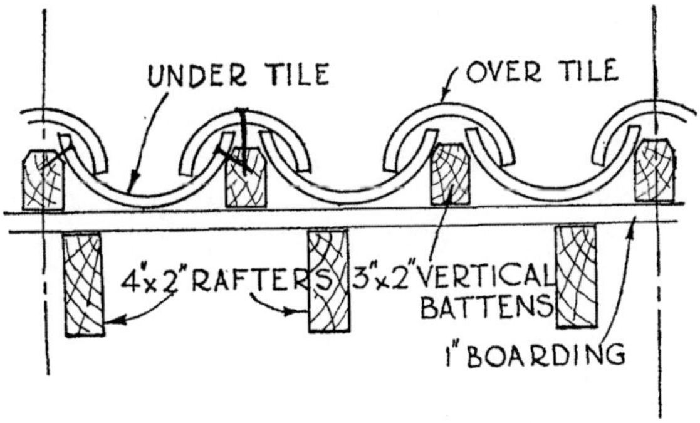

Fig. 20.—Section through Spanish tiling

Abutments

The best finish at abutments is without doubt provided by soakers and lead flashings. Architects prefer, however, to have as little lead as possible visible on the roof,

Fig. 21.—VALLEY TILING

Fig. 22.—A LACED VALLEY

and the fashion has spread of having a tile flaunching or tile flashing instead of the stepped flashing. Where this form of construction is adopted the tile flashing must be let into a chase : if this is not done, and roof settlement occurs, the tile fillet will come away from the wall, and water will penetrate the gap so left.

Soakers and lead flashings are best from a constructional point of view : in all cases soakers must be inserted, whether a lead flashing, tile flashing, or cement fillet is used. Cement fillets are not recommended.

Vertical Tiling

Plain tiles are frequently used as tile hanging, partly because they give an attractive appearance to an elevation, and partly because tile hanging is the most effective waterproof covering for a wall. Tile hanging, properly carried out, is the perfect damp-resister.

For vertical tiling each tile must be secured with two nails : in fact, complete nailing should be done on all pitches over 55° Walls may be battened or the tiles may be nailed direct to the joints of brickwork. Since water runs easily off vertical tiling, the lap is not important, and a $4\frac{1}{2}$-in. gauge may be used. This is convenient where the tiles have to be fixed directly to the joints of brickwork on edge.

SINGLE-LAP TILING

The kinds of single-lap tiles are so numerous that it is not possible here to do much more than mention the general principles underlying the covering of a roof with any kind of single-lap tile.

Roof Pitches

For single-lap tiling, roof pitches in general may be slower than for plain tiling, owing to the larger size of the units and the more effective lapping at side and head. A minimum of 35° is recommended, even though in practice slower pitches are seen.

Fig. 24.—COURTRAI-DU-NORD TILING ON OPEN BATTENING

Showing method adopted to ensure that the rolls of the tiles run in straight lines from eaves to ridge. (*Langley London Ltd.*)

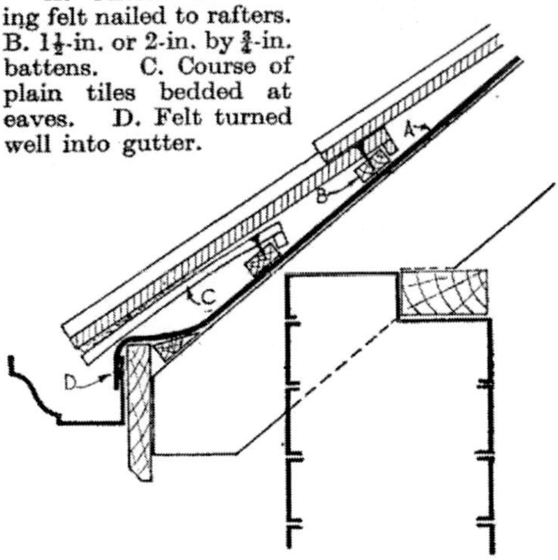

Fig. 23.—CONSTRUCTION OF EAVES IN SINGLE-LAP TILING

A. Untearable roofing felt nailed to rafters. B. 1½-in. or 2-in. by ¾-in. battens. C. Course of plain tiles bedded at eaves. D. Felt turned well into gutter.

Roof Preparation

Nearly all forms of single-lap tile require a continuous covering—boards or felt—between them and the rafters.

The ideal roof preparation is that recommended for plain

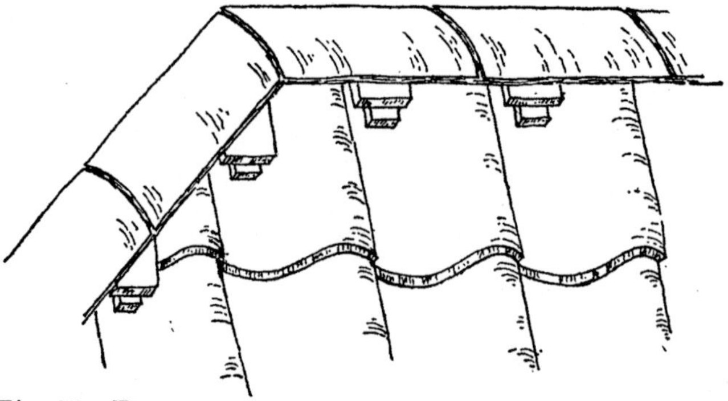

Fig. 25.—TILE LAMINATIONS FITTED BETWEEN ROLLS AT RIDGES AND HIPS

tiling, namely, plain boards, felt, counter-battens, and battens. When for reasons of cost this is impossible, then a satisfactory and reasonably inexpensive substitute is provided by untearable felt and battens. The felt should be allowed to sag slightly, but only slightly, between the rafters, and must always be carried well over the fascia and turned into the gutter. The sagging provides a series of channels down which any water that may penetrate the joints of the tiles can run to the eaves.

For Spanish and Italian tiling special preparation is necessary. The size and spacing of the vertical battens vary in different makes, and the maker's instructions must be followed. Fig. 20 is a typical section through a roof covered with Spanish tiles. The under tiles are nailed to the sides of the vertical battens, and the upper tile is fixed by a nail driven through the hole provided at the head of the tile.

Eaves

The general construction at eaves in single-lap tiling is illustrated in Fig. 23. A course of plain tiles is usually fixed over the fascia, and the first course of tiles bedded on it. No course of plain tiles is fitted at the eaves of the roof shown in Fig. 24. The tile used, a Courtrai-du-Nord, has only a small roll, and a good length of flat surface to rest on the fascia. With a pantile, or any other large-roll tile, the course of plain tiles is essential.

Verges

An undercloak should always be fitted at verges : this is even more necessary than in plain tiling. With certain makes of single-lap tile, double-roll tiles are provided for use at left-hand verges. These have no constructional value, but the appearance of the verge is improved by their use. Unfortunately, the length of roof is not always such that a double-roll tile can be used and give the correct overhang at the verge.

Ridges and Hips

The normal type of ridge and hip covering is the half-round ridge. Where a single-lap tile that shows a series of rolls and deep channels is

Fig. 26.—FOUR STAGES IN THE COVERING OF A HIP IN SINGLE-LAP TILING
(*Langley London Ltd.*)

used (for example, a pantile, or Spanish or Italian tile), tile laminations are inserted at ridges and hips to avoid the showing of large masses of mortar (Fig. 25). With pantiles only two laminations are possible, but with Spanish and Italian tiling three or more may be used. Fig. 26 shows the four stages in the completion of a hip in single-lap tiling. Note that the felt is well lapped at the hip, that it is turned over the fascia so as to come well into the gutter, and that battens run parallel to the hip on each side to give a good fixing for the splay-cut tiles.

Valleys

Where single-lap tiles are used it is better to avoid valleys altogether. The single-lap tile does, in fact, appear at its best on large four-hipped or

Fig. 27 (*Left*).—VALLEY CONSTRUCTION IN IMPROVED INTERLOCKING PANTILES —FIRST STAGE

Valley boards fixed. (*Langley London Ltd.*)

Fig. 27A (*Right*).—VALLEY CONSTRUCTION IN IMPROVED INTERLOCKING PANTILES —SECOND STAGE

Untearable roofing felt nailed to rafters. (*Langley London Ltd.*)

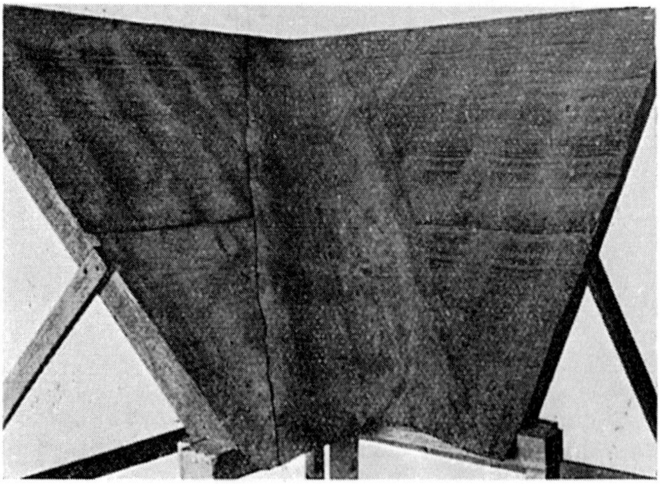

Fig. 27B (*Left*).—VALLEY CONSTRUCTION IN IMPROVED INTERLOCKING PANTILES —THIRD STAGE

Felting in valley completed. (*Langley London Ltd.*)

Fig. 27C.—VALLEY CONSTRUCTION IN IMPROVED INTERLOCKING PANTILES — FOURTH STAGE

Battens fixed on felt and lead valley fitted. (*Langley London Ltd.*)

plain-gabled roofs. Complicated roof plans, and roofs having dormers, are best covered with plain tiles.

Valleys in single-lap tiling are usually of metal, lead being the best material to use. Fig. 27 shows stage by stage the construction of a valley in improved interlocking pantiles.

A good craftsman should be employed for valleys in single-lap tiling: nothing in tiling is more difficult than to make such valleys watertight.

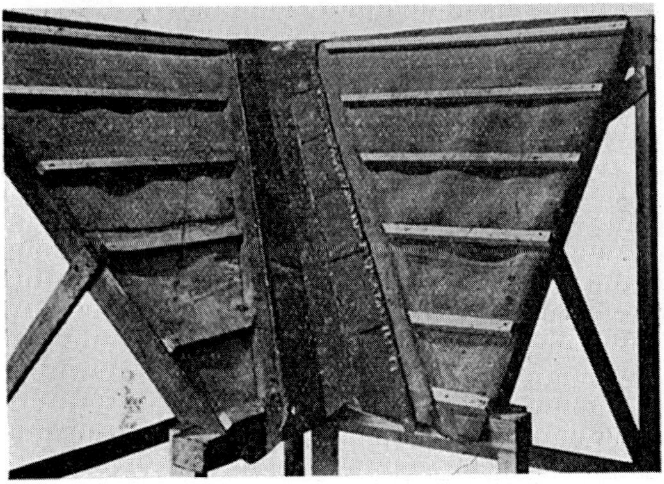

Fig. 27D.—VALLEY CONSTRUCTION IN IMPROVED INTERLOCKING PANTILES—FIFTH STAGE

Battening to valley completed. Slate undercloak fitted to valley. (*Langley London Ltd.*)

Fig. 27E.—Stage 6—tiling to valley
Note precaution to keep edges of valley straight. (*Langley London Ltd.*)

Fig. 27F.—Stage 7—valley nearing completion—straight-edge removed
(*Langley London Ltd.*)

Fig. 27G.—VALLEY CONSTRUCTION IN IMPROVED INTERLOCKING PANTILES—THE COMPLETED VALLEY. (*Langley London Ltd.*)

The photographs in Figs. 27 to 27G illustrate the various step-by-step stages in tiling a valley.

Chapter VII

ROOFING WITH ASBESTOS-CEMENT MATERIALS

ASBESTOS-CEMENT is a non-permeable, strong, fire-resisting, durable concrete, composed of asbestos fibre and cement; the asbestos fibre, unlike most aggregates, adds to the strength of the cement.

The fineness and fibrous character of the asbestos render asbestos-cement easily moulded into very thin sheets of large size, making it eminently suitable for roofing slates, tiles, and corrugated and flat sheets.

Roofing Materials

The standard forms of asbestos-cement roofing materials are :—

Diagonal or diamond-pattern slates 15¾ in. by 15¾ in.
Straight-cover slates.
Small-section corrugated sheets.
Large-section corrugated sheets.
Angular-section corrugated sheets.
Pantiles.

The dimensions and specifications of each of these types have been standardised by the British Standards Institution in *Specification No.* 690–1936, *Asbestos-cement Sheets and Unreinforced Flat Sheets and Corrugated Sheets.*

ASBESTOS-CEMENT SLATES

Diagonal-pattern Slates

Asbestos-cement as a roofing material was first known in this form. Its appearance on the roof and method of fixing are shown in Fig. 1, where also the terms used in connection with diagonal slating are illustrated. It should be noted that the lap is the amount by which one slate overlaps the ones beneath, measured at right angles to the top edge of the slate that is covered.

The slates have to be specially holed by the makers to give the desired lap—2¾, 3, 3½, or 4 in. Each slate is secured with one copper disk rivet and two slate nails; copper slate nails are recommended, particularly in acid-laden atmospheres. The slates may be laid on battens or close boards. If the latter method is adopted, care is necessary in the coursing of the slates: true lines and uniform lap are essential to the avoidance of leaks.

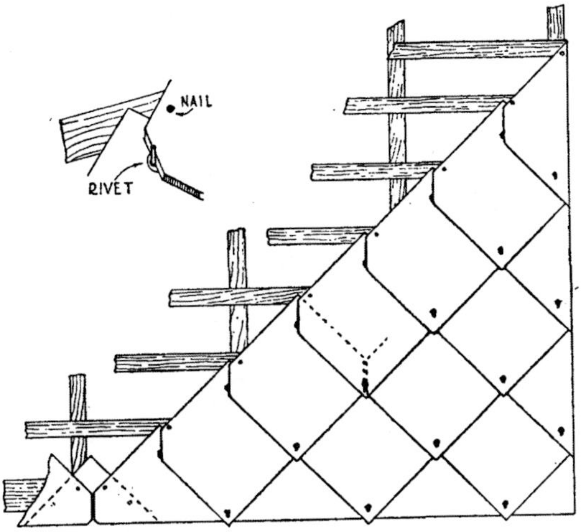

Fig. 1.—DIAGONAL-PATTERN ASBESTOS-CEMENT SLATES, 15¾ IN. BY 15¾ IN.

A variation of the diagonal pattern is the honeycomb-pattern slate, Fig. 2, in which the "drop-point" is eliminated. The method of fixing is similar to that of diagonal-pattern slates, but the honeycomb type is made only for 2¾-in. and 3½-in. laps.

Technical data for these and all other asbestos-cement materials are given in detail by the manufacturers, and the figures given in these pages have been taken from the catalogues of Messrs. Turner's Asbestos Co. Ltd., Asbestos House, Southwark St., S.E.1, and the Universal Asbestos Manufacturing Co. Ltd., 43–46 Southampton Buildings, Chancery Lane, W.C.2.

Other large makers are The Atlas Stone Co. Ltd., 15 Victoria Street,

TECHNICAL DATA FOR 15¾-IN. BY 15¾-IN. DIAGONAL AND HONEYCOMB SLATING

Method of Laying	*Lap in In.*	*No. of Slates per Square*	*Batten Gauge in In.*	*Feet of Battens per Square*	*Weight of Slates per Square in Lb.*
Honeycomb	2¾	86	8⅛	149	224
"	3½	99	7⅜	164	257
Diamond	2¾	86	8⅞	136	250
"	3	89	8½	142	261
"	3½	96	8¼	146	282
"	4	105	7⅝	158	308

These figures are net and include no allowance for waste. The weights vary slightly as between different makes.

Westminster, S.W.1, makers of "Atlas" products; and the Cellactite & British Uralite Ltd., Higham, Rochester.

Messrs. Turners are the manufacturers of "Poilite" diagonal, honeycomb, and straight-cover slates and pantiles; "Everite" corrugated sheets, rain-water goods, etc., and of Turner's Trafford Tiling, or "T.T.T." as it is popularly called.

The Universal Asbestos Manufacturing Co. are the makers of "Handcraft" corrugated sheets; slates of all three of the types mentioned above; pan tiles; "Watford" Handcraft tiles, and other materials.

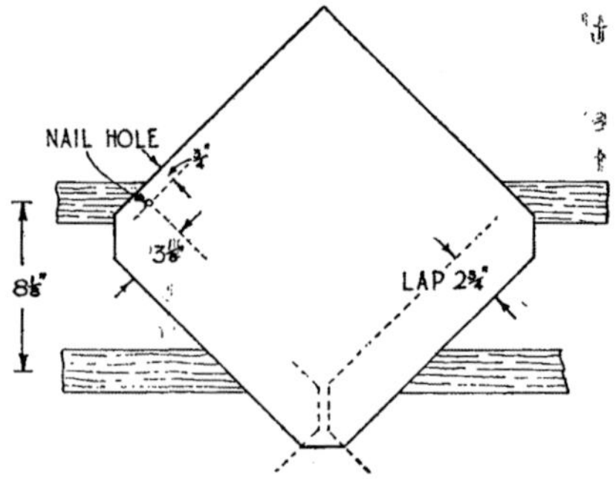

Fig. 2.—HONEYCOMB-PATTERN ASBESTOS-CEMENT SLATE, 15¾ IN. BY 15¾ IN.

Showing dimensions required by British Standard Specification No. 690–1936.

The roof pitches and their respective recommended minimum laps are as under:—

Pitch of Roof	*Minimum Lap*
40°	2¾ in.
35°	3 in.
30°	3½ in.
25°	4 in.

Straight-cover Slates

These are made in sizes 24 in. by 12 in.; 20 in. by 10 in.; 15¾ in. by 7⅞ in.; and 11¾ in. by 5⅞ in. They are laid in the same way as ordinary, natural slates, with the addition of one copper disk rivet near the tail of each slate, except in the case of the 11¾-in. by 5⅞-in. slate, which does not need a rivet. The slates are holed by the makers to give the correct lap.

TECHNICAL DATA

Size of Slate in In.	*Lap in In.*	*No. of Slates per Square*	*Spacing of Battens in In.*	*Linear Feet of Battens per Square*	*Weight of Slates per Square in Lbs.*
24 by 12	3	115	10½	114	398
24 by 12	4	121	10	119	418
20 by 10	3	170	8½	141	423
15¾ by 7⅞	3	289	6⅜	189	404
11¾ by 5⅞	2¾	519	4⅝	260	422

These figures are net, and include no allowance for waste. The weights vary slightly as between different makes.

CORRUGATED SHEETS

Two kinds of corrugated sheets are manufactured—the small section and the large. Sheets produced by any one maker are not necessarily identical in all respects with those made by any other firm, but the information given below may be taken as fairly representative of the products of all firms.

The small, or standard, section corrugated sheet is stocked in 4-ft. to 10-ft. sheets, in lengths which are multiples of 6 in. The standard width is 30 in., the pitch of corrugations is $2\frac{7}{8}$ in., the over-all depth of corrugations is $2\frac{7}{8}$ in., the reputed thickness of the sheets is $\frac{1}{4}$ in.

The end lap recommended for pitches down to about 30° is 6 in. For lower pitches, or in very exposed positions, the end lap should be increased or sealed with mastic or other suitable material. With a 6-in. lap and a side lap of $1\frac{1}{2}$ corrugations, the actual cover of a 9-ft. sheet is 8 ft. 6 in. by 2 ft. $1\frac{1}{2}$ in. The weight of 100 sq. ft. as laid is 270 lbs. to 330 lbs., according to the make and size of sheet. (The larger the sheets, the smaller the area that is covered twice.)

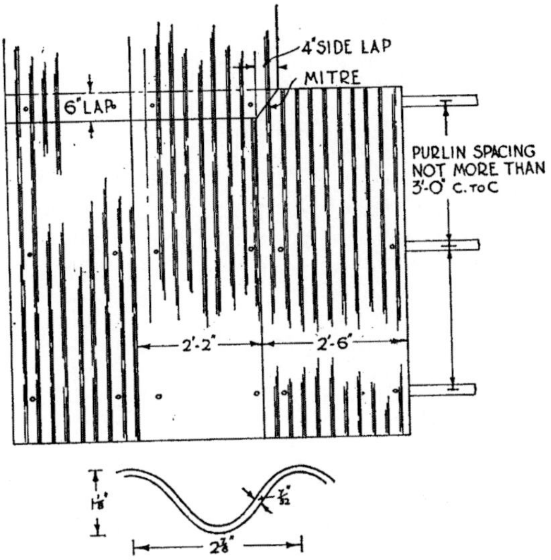

Fig. 3.—STANDARD-SECTION ASBESTOS-CEMENT CORRUGATED SHEETS FIXED TO WOOD PURLINS

Dimensions required by Specification No. 690–1936.

Fixing Corrugated Sheets

Fig. 3 shows in plan the method of fixing to wood purlins. Three-inch by $\frac{1}{4}$-in. galvanised iron screws, with lead cupped and fibre washers, are used, and it is important that the side laps should not be bolted together; nor should the sheets be fixed more tightly than is necessary to attach them securely. The purlin spacing should not exceed 3 ft. Detailed instructions for fixing are issued by each manufacturer.

For fixing to iron purlins, galvanised hook bolts are supplied by the makers.

The large-section corrugated sheet, of which the Everite "Big-Six" and Handcraft "Super-Six" are typical examples, differs from the standard sheet in that it has larger corrugations and is slightly thicker. It is stocked in 4-ft. to 10-ft. lengths, in lengths which are multiples of 6 in. The width varies between 40 in. and 43 in., the pitch of corrugations is 5¾ in., and the depth 2 in. Purlins may be spaced up to 4 ft. 6 in., as compared with the maximum of 3 ft. with standard sheets. The weight per square as laid varies between 310 lbs. and 350 lbs. The general fixing details are as for standard sheets, but the makers' instructions must in all cases be strictly followed.

Angular-section Corrugated Sheets

Angular-section corrugated sheets were originally designed to provide a rather better-looking roof than is given by the ordinary corrugated sheet. A typical form is illustrated in Fig. 4, but the sheets made by various makers differ in details. Fig. 4 gives the dimensions for an angular-section corrugated sheet as demanded by the British standard specification, a tolerance of one-quarter of 1 per cent. being allowed.

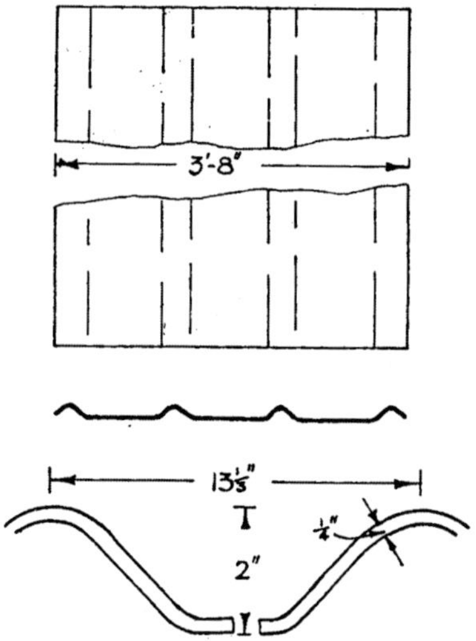

Fig. 4.—TYPICAL FORM OF ANGULAR-SECTION CORRUGATED ASBESTOS-CEMENT SHEET

PANTILES

The asbestos-cement pantile has been designed on the lines of the English pattern of clay pantiles. It has certain advantages over the latter: it is lighter in weight; more regular in shape, and therefore fits more closely on the roof; and it can be laid on pitches down to about 40° without felt or boarding.

Technical Data

	Large Pantile	*Small Pantile*
Size	15¾ in. by 13¼ in.	15¾ in. by 9$\frac{7}{16}$ in.
Standard lap	4 in.	4 in.
No. of tiles per square (100 sq. ft.)	108	164
Centres of battens	11¾ in.	11¾ in.
Lin. ft. of battens per square (100 sq. ft.)	102	102
Approx. weight of tiles per square (100 sq. ft.)	362 lbs.	392 lbs.
Approx. weight of 1,000 tiles	29 cwts.	21 cwts.
Add 5 per cent. for cuttings to valleys, hips, etc.		
Size of battens	2 in. by 1 in.	
Colour	Russet-brown	
Pitch of roof	Not less than 40° when laid on open battens. When boarding and/or felt is used under pantiles, a pitch of 30° can be adopted.	

Method of Fixing

Fix pantiles with two galvanised nails per tile (one 2 in. by 10's gauge in the right-hand hole, and one 1½ in. by 10's gauge in the left), driven

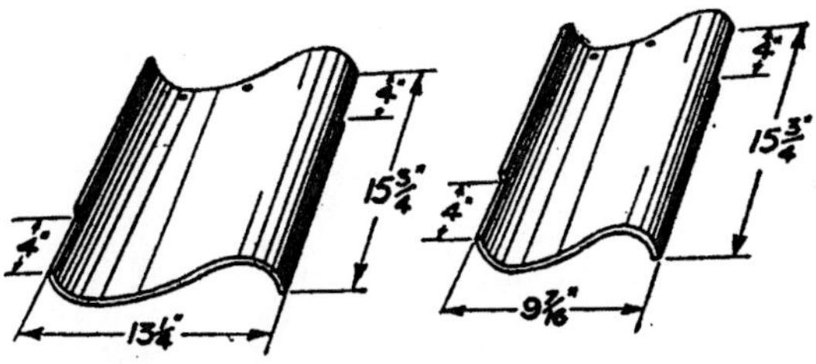

Fig. 5.—Asbestos-cement pantiles

at right angles to the surface of that portion of the tile through which they pass.

A proved method is to fix nail A first, 2 in. long, but driven only partly home. Next, drive nail B, 1½ in. long, fully home. Finally, return to nail A, which is just off that portion of the tile bearing on the batten, and drive it far enough to cause the roll of the tile to bind on the next tile under, as shown at point C.

There will be a tendency for the vertical lines of tiles up the slope to lean to the right, and to avoid this each pantile should be held firmly to the left while it is being nailed.

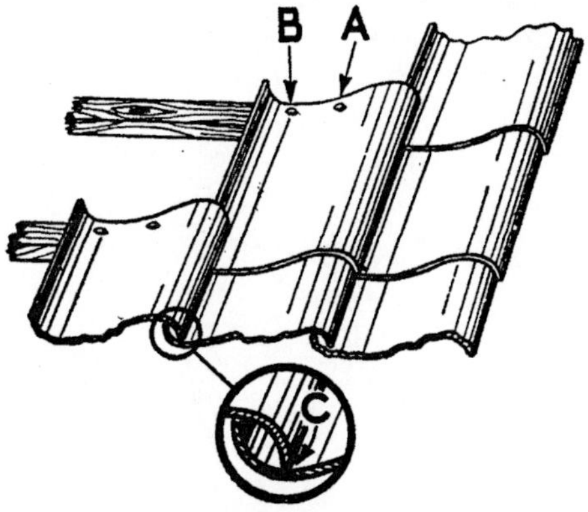

Fig. 6.—Fixing pantiles

Eaves Finish

Stop-end pantiles are supplied, only in the large size, for use at eaves. These tiles are made with a closed end, which, when used in conjunction with an eaves strip, eliminates the gap over the back of the gutter.

In the case of the small pantiles, the eaves pantiles should be bedded on an asbestos-cement strip with expanded metal, nailed or stapled in position, used to assist keying. The ends of the pantiles should afterwards be pointed in cement.

Hips, Valleys, Verges, and Ridges

Twin pantiles have been specifically designed in both small and

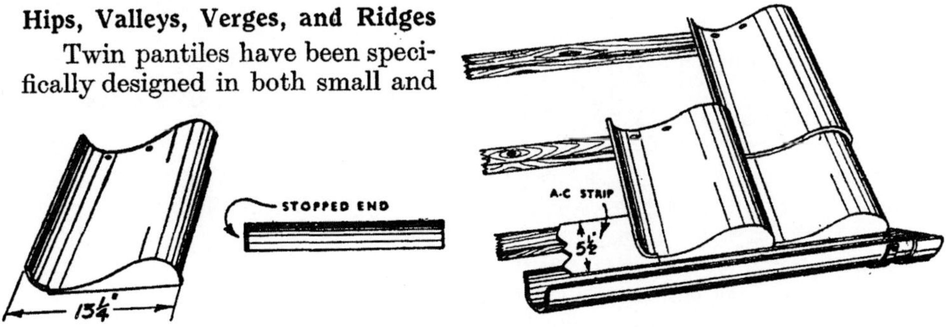

Fig. 7.—Eaves finish for large pantile roof

large sizes to avoid small triangular pieces of tile which would otherwise occur at hips and valleys, and be difficult to fix securely. The above shows a typical instance where these twin pantiles are used. The illustrations below show the large-pattern twin pantile.

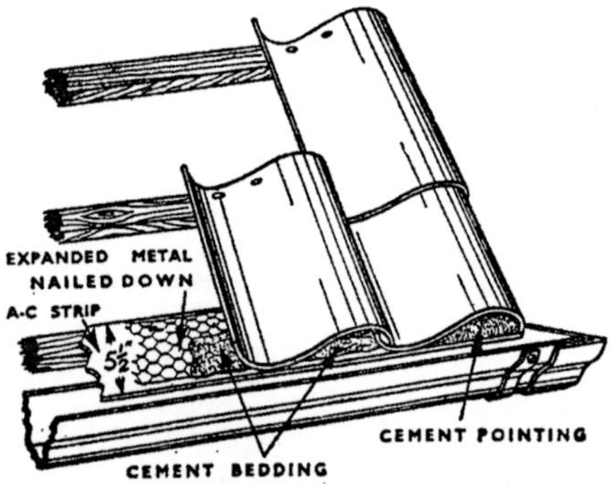

Fig. 8.—Eaves finish for small pantile roof

Double-roll pantiles, made in both small and large sizes, are for use on left-hand verges, and enable a roll finish to be given to a roof verge.

There are two ridges for accommodating varying angles of slope. No. 1 (Fig. 10) is generally used for ridge capping, and No. 2 pattern for hip capping. Ridges are lined with concrete to give a good key to the cement bedding, by which means the ridges should be fixed, afterwards

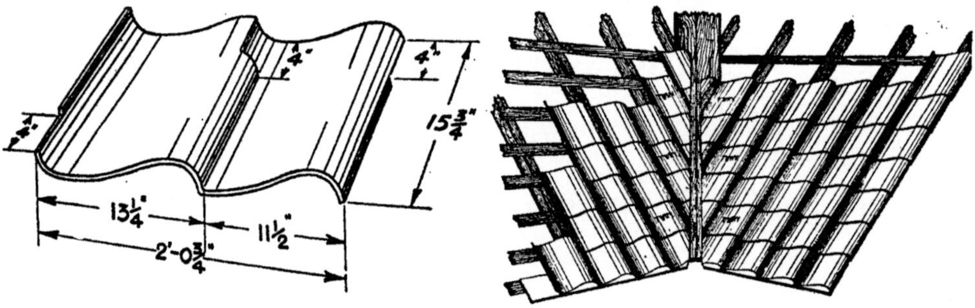

Fig. 9.—Twin pantile and its application at valley

being pointed to give a clean finish. Hip irons should always be used.

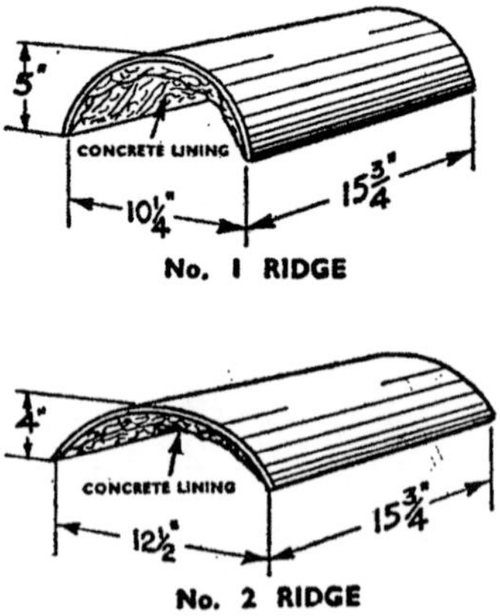

Fig. 10.—Ridges

Bedding

The bedding materials for pantiles and ridges should consist of cement or hair mortar, and must be used fresh.

It is recommended that at verges and valleys asbestos-cement strips and expanded metal be fixed, and the pantiles be bedded in cement on this strip, afterwards pointing up with cement, to obtain a flashing; this method obviates the work of dressing lead to the contour of the under side of the pantiles at these points.

Chapter VIII

COVERING ROOFS WITH ASPHALTE SHINGLES

ASPHALTE as used for roofing is well known for its durability, being impervious to water and providing an effective barrier to the passage of heat and cold. Its use as a covering for flat roofs has already been mentioned in Chapter III.

A form of asphalte shingle has been developed as a covering for sloping roofs and placed on the market under the name of Flextile ; this material consists of high-quality asphalte roofing covered with a layer of slate or tile chippings on the exposed surface. The roofing is cut into strips 36 in. × 10 in., these strips being further subdivided into four by a slot or " cut-out " 4 in. deep. The strips are laid with 4 in. exposed to the weather, giving a 6-in. overlap and 2-in. cover as in tiling. The appearance of the finished job resembles that of a slated or tiled roof, the " cut-outs " throwing a sufficient shadow to give an appearance of solidity. The shingles are made in two colours, rust-red and grey-green ; the colours are natural, since they are due to the chippings applied to the asphalte roofing, and weather similarly to slates or tiles.

The shingles are light, weighing only 200 lbs. per square, thus assisting economy in roof design. The light weight reduces the handling of the material and enables the shingles to be easily and quickly laid.

This type of roofing lends itself easily to awkward shapes, since the shingles can be cut with a knife to fit round dormers, curved soffits, valleys, etc.

Laying of Flextile Shingles

The shingles should be applied on top of close-jointed boarding, and the valleys and gutters should be constructed first. The valleys and gutters may be of lead or copper, but if desired may be formed with asphalte roofing strips in two layers. The roofing strips are supplied for the purpose in any width required, the recommended widths being 12 in. for the under strip and 18 in. for the upper, thus bringing the edge well under the shingles, as shown in Fig. 1.

The understrip is nailed to the boarding, and the upper strip cemented to it with mastic applied hot. The upper strip is nailed at the edges only, as shown in Fig. 1, thus leaving no nail heads exposed when the roof is completed.

A strip of asphalte roofing is next fixed at the verges as shown in Figs. 2

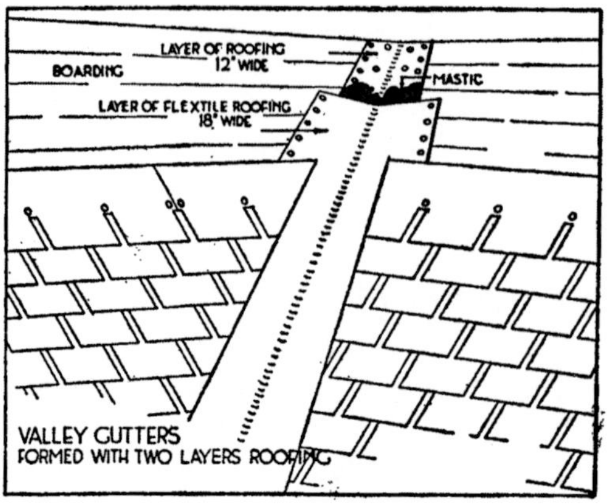

Fig. 1.—Showing double protection of Flextile roofing strips in valley (*D. Anderson & Son Ltd.*)

and 3. The eaves course should be a row of shingles reversed (see Fig 2), or alternatively a strip of asphalte roofing not less than 9 in. and preferably 18 in. wide may be used as shown in Fig. 3.

The laying of the shingles then proceeds in the same manner as slating or tiling, starting at the eaves and working up the roof. Fig. 4 shows the first row of shingles in position. Each alternate row should begin by cutting off half of the first leaf, so as to bring the joints in each row midway between those in the next.

The first course proper should overhang the edge of the roof boarding by about $\frac{3}{4}$ in., with a similar overhang at the verges. The asphalte roofing strips should be bent over and nailed to the edges of the boarding as shown in the illustrations. If barge boards are used, the shingles are finished flush with the edge of the barge boards.

Lap

The width of the shingles being 10 in. and the depth of the slot 4 in., they are laid so that the bottom edge is in line with the top of the slot in the course below, thus giving the 6-in. overlap already described.

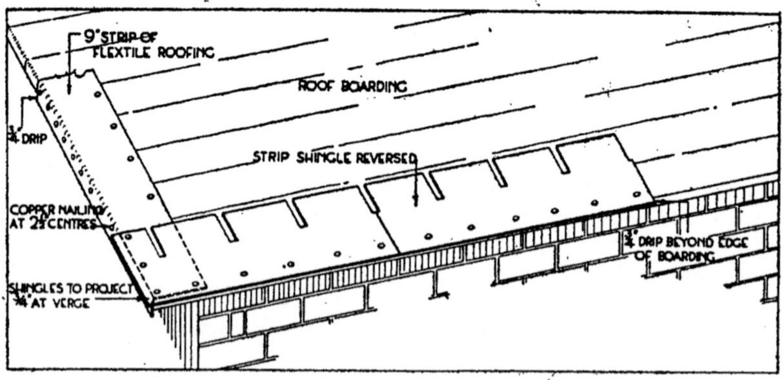

Fig. 2.—Showing 9-in. strip of Flextile roofing at verge and starting or eaves course of shingles reversed

This course is begun by cutting off half of the first leaf to ensure breaking joint when the first course proper is laid. (*D. Anderson & Son Ltd.*)

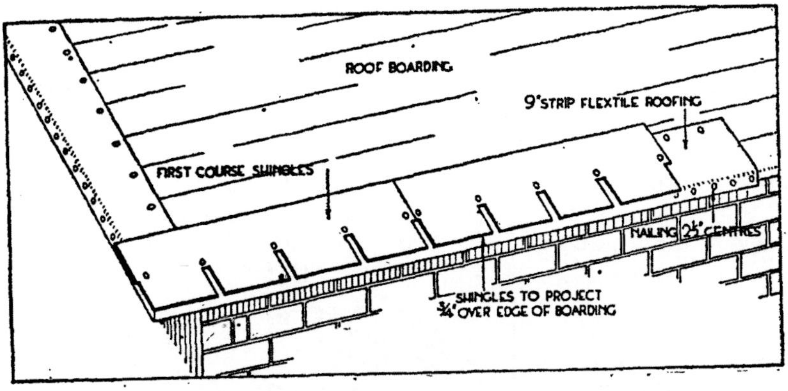

Fig. 3.—Showing strip of Flextile roofing as starting or eaves course with first course of shingles in position
Note the Flextile roofing bent over and nailed to edge of boarding.
(*D. Anderson & Son Ltd.*)

Nailing

One-in. galvanised clout nails are used, each course being nailed ½ in. above each slot as shown in Figs. 3 and 4. At projecting edges exposed to wind pressure some additional nailing should be provided, and care should be taken to avoid driving nails between the joints of the boarding.

Ridges and Hips

There are two alternative treatments for ridges and hips in this type of roofing. The first is to use asphalte roofing strips 9 in. wide, bent over the ridge or hip, lying 4½ in. on each side, and nailed along the edge at 2-in. centres ¾ in. from the edge as illustrated in Fig. 5.

Alternatively. the shingle strips may be cut into individual pieces

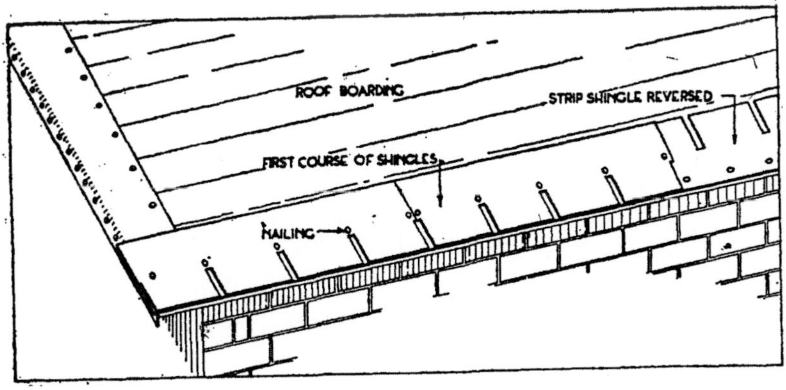

Fig. 4.—Showing first row of shingles in position. (*D. Anderson & Son Ltd.*)

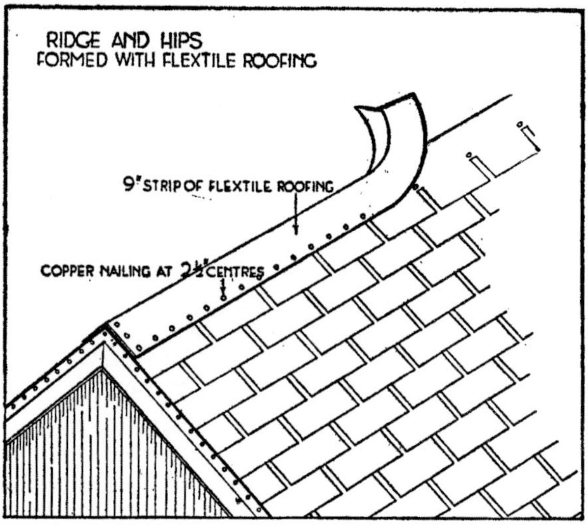

Fig. 5.—SHOWING ROOF BEING CLOSED IN AND COMPLETED WITH STRIP OF FLEXTILE ROOFING AT THE RIDGE (*D. Anderson & Son Ltd.*)

9 in. × 10 in., bent over the ridge or hips, overlapping to expose 5 in. to the weather and nailed as shown in Fig. 6.

Flashings

Flashings are carried out in the usual manner employed in slating. Upstands may be formed of asphalte roofing strips if desired, but metal is recommended for all counter flashings.

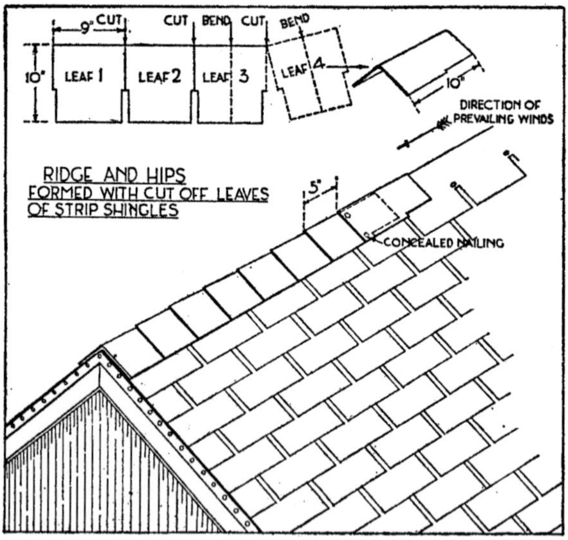

Fig. 6.—SHOWING METHOD OF CUTTING STRIP IN PIECES AND FIXING OVER THE RIDGE (*D. Anderson & Son Ltd.*)

The constructional details and technical data relating to the use of asphalte shingles which are given in this chapter have been based on information placed at our disposal by Messrs. D. Anderson & Son Ltd., to whom we have pleasure in making this acknowledgment.

Chapter IX

COVERING ROOFS WITH ASBESTOS-FELT

THE following information relates to the use of asbestos-felt roofing material.

STANDARD WEIGHTS AND MEASUREMENTS OF MATERIAL

Material	*Length of Roll*	*Width*	*Weight per Roll*
Asbestos-felt, standard weight (uncoated)	24 yds.	1 yd.	30 lbs.
Asbestos-felt, No. 1 ply coated	12 yds.	1 yd.	36 lbs.
Asbestos-felt, No. 2 ply uncoated	12 yds.	1 yd.	33 lbs.
Asbestos-felt, No. 2 ply coated	12 yds.	1 yd.	56 lbs.
Asbestos-felt (fire-resisting)	12 yds.	1 yd.	63 lbs.

Laying Felt

For convenience in handling, cut rolls into 18-ft. or 24-ft. lengths, stretch felt smoothly and evenly next to and along eaves, allowing it

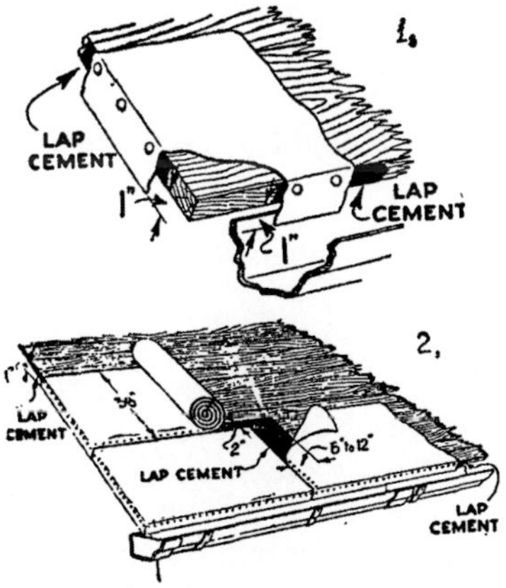

Figs. 1 *and* 2.—LAYING FELT

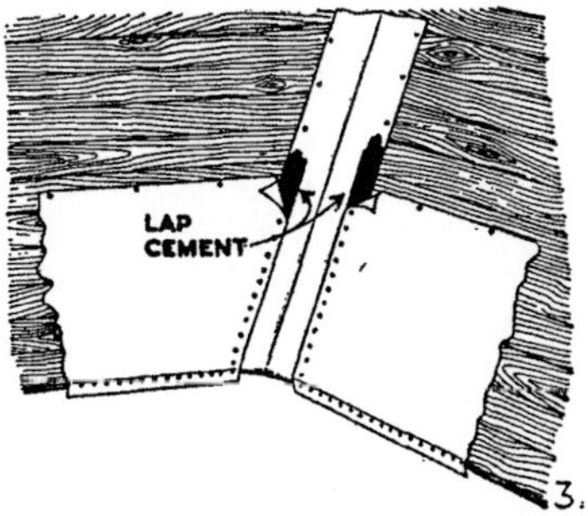

Fig. 3.—LAYING FELT AT VALLEY GUTTER

to project so that when turned down it will hang about 1 in. below edge of roof boards at eaves and gables, then drive a few clout nails about ½ in. from top edge to fasten temporarily. Lower edge of second piece should overlap top of first piece by 2 in. Secure temporarily along top edge as before. Raise lower edge of second sheet, and apply between this lap a thorough coat of bituminous solution or lap cement. Allow lower edge of upper sheet to fall back and embed it thoroughly in jointing cement. Side laps should be laid in a similar manner, but with a cover of from 6 in. to 12 in., and arranged in staggered formation. For nailing, use ordinary clout nails, ½ in. long when working on ¾-in. boards, and ¾ in. long when working on 1-in. or thicker boards. Drive nails through centre of lap 2½ in. to 3 in. apart. Coat all nail heads and exposed edges with jointing cement.

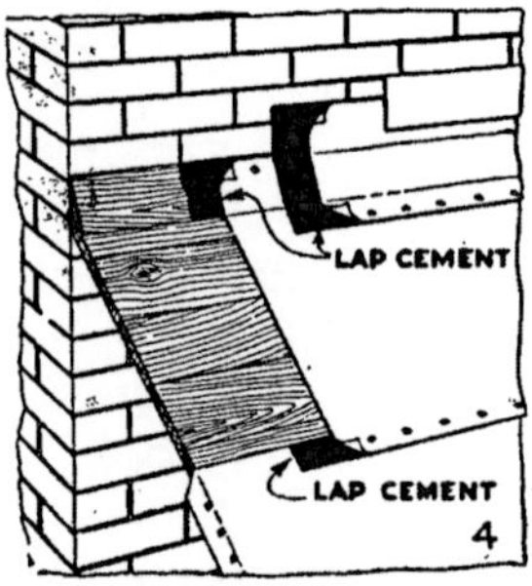

Fig. 4.—FINISHING AT LEAN-TO

Felt projecting over eaves and verges should now be turned down (edges of boards being coated with lap cement), and securely nailed.

Valley Gutters

(See Fig. 3.) Strip of roofing felt should be wide enough to underlap felting at sides by not less than 6 in. This lap should be cemented, and nailed every 2½ in. through the lap, at ¾ in. from the exposed edges.

Lap Cement

(11 gals. = 1 cwt.) Approximate quantity required for laps per 12-yd. roll is ½ pint.

Clout Nails

For every 12 yds., 100 nails should be allowed when used for sarking, or 300 nails when used for roofing purposes.

Flashings

Make vertical flashings by bending felting up about 4 in. against wall. Thoroughly cement turned-up part to wall, and nail at upper edges with nails about 9 in. apart. Fit strips of roofing felt 12 in. wide length-

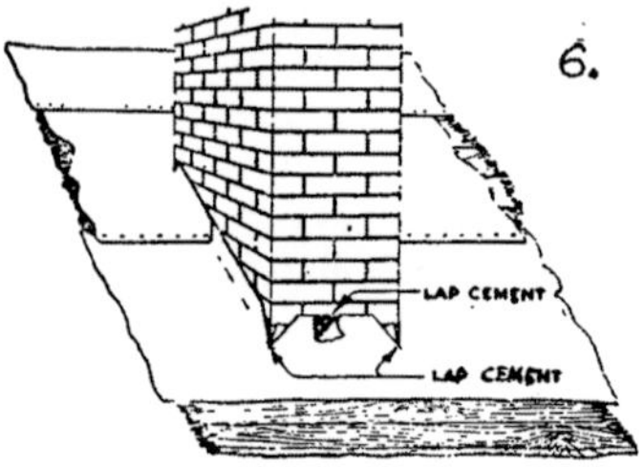

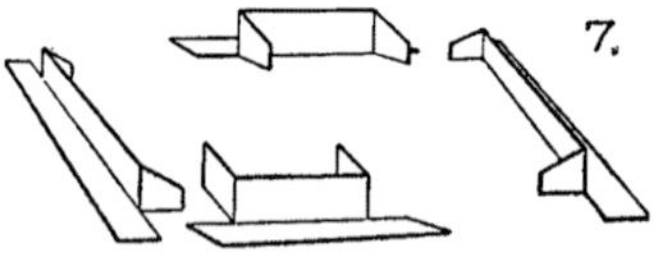

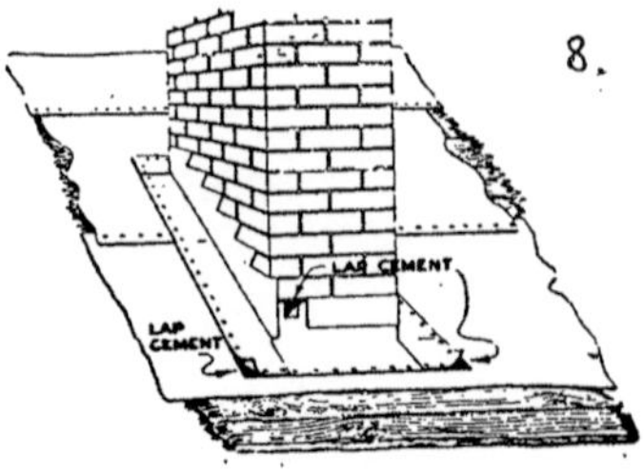

Figs. 5 *to* 8.—FLASHINGS AT CHIMNEY

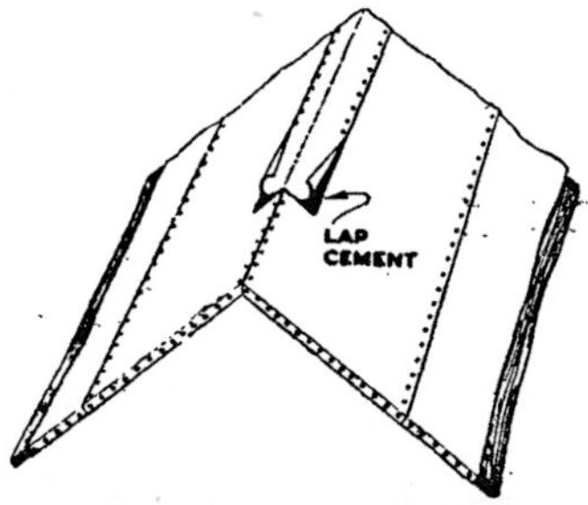

Fig. 9.—Ridge treatment

wise into angle over the turned-up edge, and cement to wall and roofing underneath. Nail lower edge of strip every 2½ in. through cemented lap, about ¾ in. from bottom exposed edge. Upper edge of this strip may be flashed with a lead apron flashing, or, as an alternative where there is a coping, a strip of roofing felt may be cut, wide enough to extend under the coping and over the felting to roof level, where it should be cemented and nailed.

Flashings at Chimneys

Cut pieces of felt 4 in. square and fit them around corners of chimneys, as shown (Fig. 5). Then lay roofing in manner already described, cutting it so that it may be turned up, fitted, and cemented round chimney (Fig. 6) about 4 in. For each side of chimney, cut strip of roofing 12 in. wide, and 12 in. longer than side to be flashed. Also cut pieces to fit above and below chimney. Cut and fold as shown in diagram (Fig. 7) and thoroughly cement to chimney and roofing underneath. Step and apron flash the upper edge as shown in Fig. 8, the felt being stuck down with lap cement, and sealed into brick joints with bituminous cement.

Ridges and Hips

Use strip of roofing felt at least 9 in. wide, bending it over ridge as a capping. Thoroughly cement between laps, and nail as already described.

Sarking

For convenience in handling, when fixing asbestos-felt for sarking under slates, tiles, or corrugated sheeting, cut roll into 18-ft. or 24-ft. lengths. Lap 2 in. Use similar nails as for roofing. Drive nails through centre of lap, at about 9 in. apart.

If the sarking felt is to be applied over spars and under slate laths, it should be evenly stretched in position over spars and temporarily secured, allowing 2-in. overlap (see lower illustration). Fixing laths over felt will secure it. Where possible, it is advisable to nail felt to under side of laths, through the lap between the spars.

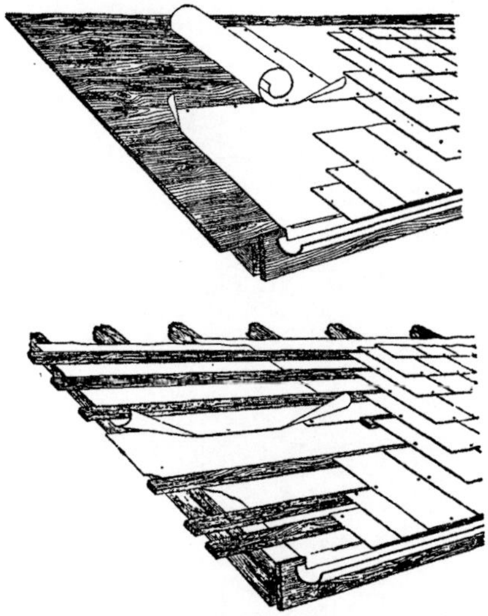

Fig. 10.—Sarking

Chapter X

COVERING ROOFS WITH WOOD SHINGLES

WOOD shingles are becoming increasingly popular in England, and this is not surprising, in view of their suitability for buildings in well-wooded country districts. Two objections are frequently raised when shingles are put forward as a suitable covering for domestic work : their non-durability, and their non-resistance to fire.

The durability of wood shingles is much greater than is popularly supposed ; there are many shingled roofs of oak or cedar in England and Wales that have stood for forty years and more, and are still perfectly sound. The risk of fire is also greatly exaggerated. In British Columbia, where red-cedar roofs are universal, fires that originate in the roof are exceedingly rare. If desired, the shingles can be coated with a fire-preventive paint.

Types

Shingles are commonly of English oak or imported cedar, the former being the more expensive. The photographs illustrating this chapter are of Canadian red cedar-wood shingling supplied by Messrs. W H. Colt Ltd., to whom we are also indebted for much of the following information regarding fixing.

Fig. 1.—CANADIAN RED CEDAR SHINGLES ON OPEN BATTENING AT 5-IN. GAUGE, SHOWING FORMATION OF RIDGE AND HIPS. (*W. H. Colt Ltd.*)

Size

Cedar-wood shingles are 16 in. long; $\frac{2}{5}$ in. thick at the butt, tapering to the top. The widths vary from 4 in. to 12 in. A 5-in. exposure (or gauge) is satisfactory for all pitches down to 30°; for slower pitches the gauge should be decreased.

Shingles are sold in bundles, the covering capacity being as follows :—

At $3\frac{3}{4}$-in. gauge	one bundle covers	19·2 sq. ft.
,, 5-in. ,,	,, ,, ,,	25·7 ,, ,,
,, $5\frac{1}{2}$-in. ,,	,, ,, ,,	28·3 ,, ,,
,, $7\frac{1}{2}$-in. ,,	,, ,, ,,	38·5 ,, ,,

The $7\frac{1}{2}$-in. gauge is recommended only for vertical shingling.

Weight

The weight of a square of shingling is approximately one-tenth of the weight of a square of plain tiling : an ordinary specification for a shingled roof calls for 2-in. by 3-in. rafters at 2-ft. centres, and 1-in. by 2-in. battens at 5-in. centres.

Nailing

Shingles are not holed for nailing : the nails are driven directly into the shingles at about 1 in. from the edge, and about 1 in. above the gauge line (that is to say, about 6 in. from the butt). The heads should not be driven into the wood. The nails may be ordinary galvanised slate nails, $1\frac{1}{2}$ in. long and of No. 12 or 13 gauge.

Fig. 2.—VERGES, RIDGES, ABUTMENTS, VALLEYS, AND VERTICAL WORK IN RED CEDAR SHINGLING. (*W H. Colt Ltd.*)

Laying

Fig. 1 illustrates the method of laying. A double course is laid at the eaves, as in plain tiling, projecting about 2 in. beyond the fascia. In continuing with the courses up the roof, care should be taken that a minimum side lap of $1\frac{1}{2}$ in. is given, but no attempt should be made to sort the shingles into widths. If they are used directly from the bundles, the correct random effect will be obtained. A gap of $\frac{1}{8}$ in. to $\frac{1}{4}$ in. should be allowed between adjacent shingles.

Hips and Ridges

Hips and ridges are best formed from the shingles themselves, using shingles of an even width of, say, 4 in. Wider ones may have to be cut down for this purpose. The method of butting the ridges at edges and overlapping at ends is clearly shown in Fig. 1. Lead or felt is not needed at ridges and hips. Alternatively, hips and ridges may be boarded, using timber not less than 4 in. wide.

Verges and Abutments

Wide shingles should be used in alternate courses at verges and abutments, as in Fig. 2.

Valleys

Valleys are formed with lead in the same way as for a tiled or slated roof. A good width of lead, say 15 in., should be used, with no tilting fillet.

Walls

Walls are frequently shingled in staggered courses: this variation in method may also be employed for roofs. The gauge on vertical work may be increased to $7\frac{1}{2}$ in., to which exposure the shingles on the mansard and vertical work of Fig. 2 have been fixed.

Colour

Both oak and red cedar-wood shingles improve greatly in appearance after a few months' exposure to the weather. The original red tinge of the latter changes to an attractive silver-grey. If any special colour is desired a good creosote stain should be used, but the user who dislikes the first appearance of a shingled roof is strongly recommended to wait until the roof has weathered before finally deciding whether or not to stain.

Chapter XI

EMERGENCY REPAIRS TO ROOFS

MUCH of the activity of the building trade is now taken up with the repair of houses which have sustained minor damage by air bombardment. These repairs are of an emergency and often of a temporary nature owing to shortage of materials and labour available.

Among the most vulnerable parts of a house to blast from bombs is the roof, and here it is essential that such damage be repaired promptly ; the longer such attention is delayed, the longer is the interior of the house exposed to the elements and the greater is the depreciation.

The actual roof repairs that require to be effected comprise mainly broken or stripped roofing slates or tiles. Minor holes in the roof covering are best replaced with the same material as the rest of the covering.

A roof completely stripped of its slate or tile covering but with the timbers intact can be re-covered with the same material if this is available. If not, it may be covered up with roofing felt, battened up at the edges, and given a coat of tar ; methods of covering with roofing felt, and other emergency materials, such as corrugated iron, are given later.

Inspect the Roof

Before commencing to retile or reslate, inspect the roof framing to see whether any rafters have lifted from the ridge by the force of the blast and that purlins and struts are in place and well secured. Where a rafter has been snapped by a piece of flying debris, a repair can be effected in most cases by fixing a new piece of timber to the broken one.

In cases where several roof members have collapsed, it may be necessary to remove all the covering and battens before respiking the members together again and replacing any damaged timbers.

In more drastic cases, it may be found necessary to dismantle what remains of the roof and re-erect. Consideration can be given to the possibility of replacing a pitched roof by a flat one, consisting of joists spaced fairly wide apart and covered with insulating board and roofing felt, the whole being given a coat of tar.

Repair of Slate Covering

Care has to be taken when replacing damaged slates to disturb the undamaged parts of the roof as little as possible. Old roofs which have proved satisfactory often become troublesome when disturbed. In such cases "slurrying" will prove helpful.

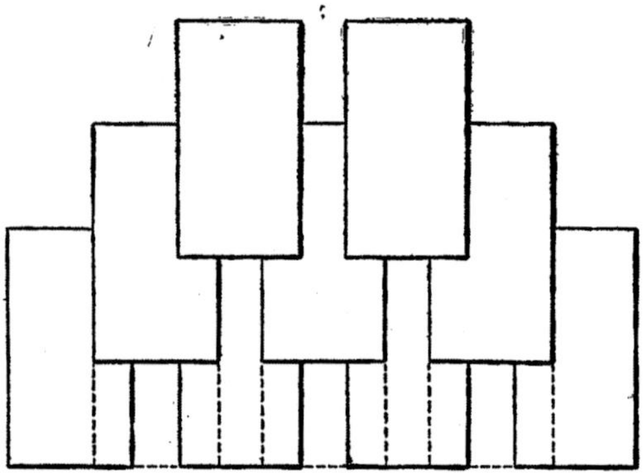

Fig. 1.—Covering roof with slates in extended order

Slates shown dotted form double course at eaves.

Another source of trouble is the difficulty in obtaining the necessary slates of the right size to effect a proper repair even if salvage can be obtained nearby. One method of overcoming this is to use whatever sizes are available on the damaged area. This will possibly necessitate rebattening and the forming of the junction between the two sections with a felt course, thus effectively breaking the bond without reducing the lap.

Another method that can be adopted to overcome the shortage of the right size of slates is to cover the roof with slates in extended order, as a temporary measure. This method will allow a saving of one-third of the normal number of slates (see Fig. 1).

Where the damage is only slight, the slates may be replaced by asbestos-cement sheeting which has been cut to the right size.

Repair of Tile Covering

Repairs of a tiled roof may have to be made with tiles of a different curvature. In order to avoid leakage, torching with mortar from inside the roof at the junction between the new and old work may have to be resorted to.

COVERING ROOFS WITH BITUMEN ROOFING FELT

There are many kinds of bitumen roof coverings on the market that are a great improvement, both in the matter of handling and laying and in weathering qualities, over the old tarred and sanded " felting."

How to Cut the Felt

Unroll the material on a level, smooth surface, where no stones, etc., will cut through. Avoid walking on the roofing when unrolling. To cut, use a straight wood batten 3 ft. long, and a sharp, pointed knife —a lino knife is excellent. Draw the knife towards you when cutting.

As a rule roofings are supplied in rolls 3 ft. wide by 12 yds. long.

Fixing Felting Up and Down

The best arrangement of fixing felting on a boarded roof is to use it with the lap joints running in a vertical direction, and when employed in this way quite flat roof pitches, not exceeding say ¼ pitch, can be dealt with successfully (Fig. 2).

The provision of rolls or ribs underneath the felting introduces the principle of weather-tightness which is a feature of the long joints in corrugated-iron roofing. Water won't run uphill. Therefore if the 3-ft. wide covering is laid on ribs 1 ft. 5 in. apart, every alternate strip forming a joint-rail, the danger of water driving through vertical joints will be avoided.

Further, the wide corrugations formed in the roof will tend to prevent or lessen the rucking of the roofing material after a spell of really hot weather.

Double Covering

For a building requiring a better construction, viz. a double thickness of bitumen felting, the arrangement sketched in Fig. 3 may be adopted, quite the cheapest material being used for the undercoat and a "2-ply" or even a superior stuff for the outer one. A detail of the finish at the gable end of a building is illustrated in Fig. 4.

Where the two thicknesses of roofing are employed the gable end would have the inner coating placed under the lifting batten, as shown in the sketch (Fig. 3).

The Battens

In all cases these battens must be chamfered or rounded on the top edges to prevent cutting the felting at the bends.

If there is any choice it is always as well to fix felting material in warm weather. There is less danger of it being cracked in the process of fitting up although of course it is softer and requires careful handling in other respects.

Undercoating

Where an undercoating is used this may be laid on with either horizontal or vertical joints, preferably the former on larger surfaces.

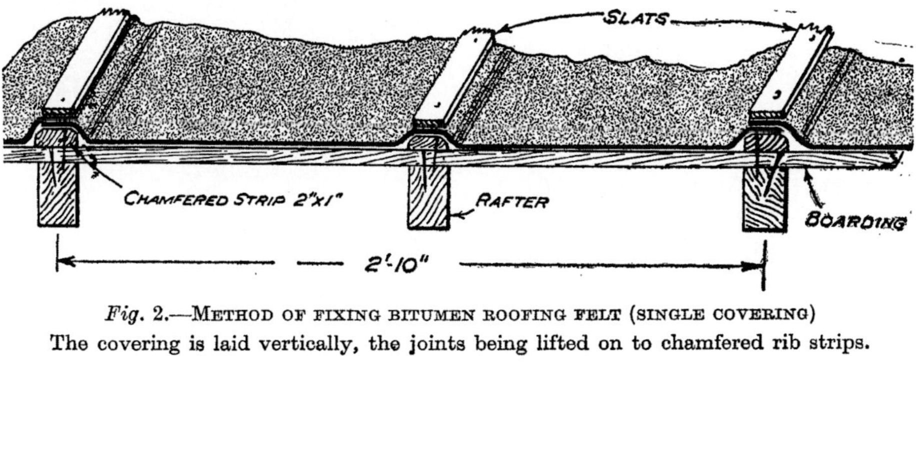

Fig. 2.—METHOD OF FIXING BITUMEN ROOFING FELT (SINGLE COVERING)
The covering is laid vertically, the joints being lifted on to chamfered rib strips.

Covering Ridges

With ridges, two methods are possible. If the ridge board stands up well above the surface of the roof as shown in Fig. 5 (A), a common constructional detail, then each roof pitch may be dealt with by a separate length of felting, the joint at the top being as shown and the whole being covered by a zinc or other ridge capping.

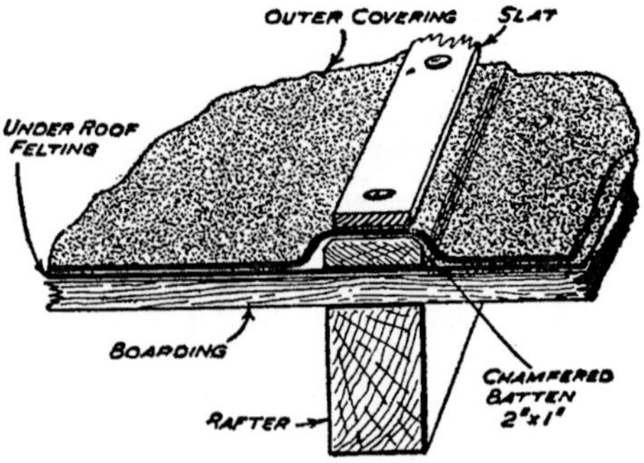

Fig. 3.—METHOD OF DOUBLE COVERING WITH ROOFING FELT

In some cases the writer has used a flush apex either by depressing the ridge board or using a half-lapped ridge joint for the top ends of the roof rafters. These two arrangements are indicated in the sketches B and C, Fig. 5. The roof felting may then pass over the ridge to the other pitch in one strip.

If the roofing is one of two coverings, the under layer should be tacked on with smaller clouts. Further, if the ridge does not lift the roofing to an amount equal to the thickness of the ribs, the latter must be chamfered off at the top.

Fixing Felting at Eaves

At the eaves of a shed the fixing of felting requires special care. If there is any "ripping off" in a gale of wind, it is here that the trouble usually starts. The roofing felt may be lapped over the metal guttering and secured by the usual slat, as indicated in Fig. 6. The slight tilt given to the covering is accomplished by planing off the top of the fascia board at an angle as shown. This raising of the bending line should be shown at A, which should be equivalent to the height of the "roll."

Where a double thickness is used the under sheeting may finish at the gutter back and the outer felting lap over, as depicted in Fig. 7.

FITTING CORRUGATED-IRON ROOFING

It is said of corrugated iron when used as roofing that it always looks like corrugated iron and nothing else. However this may

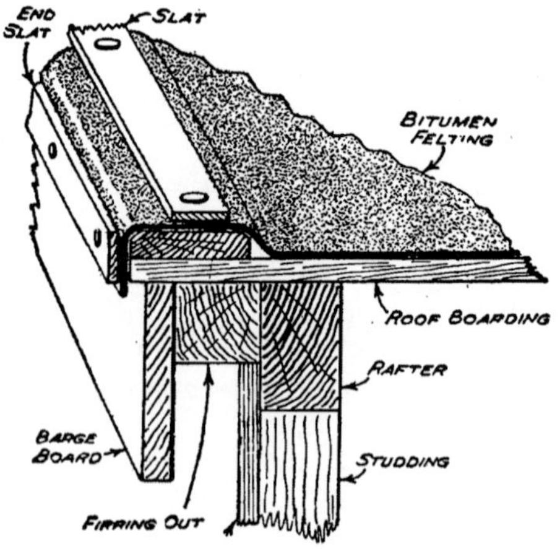

Fig. 4.—TREATMENT FOR GABLE ENDS

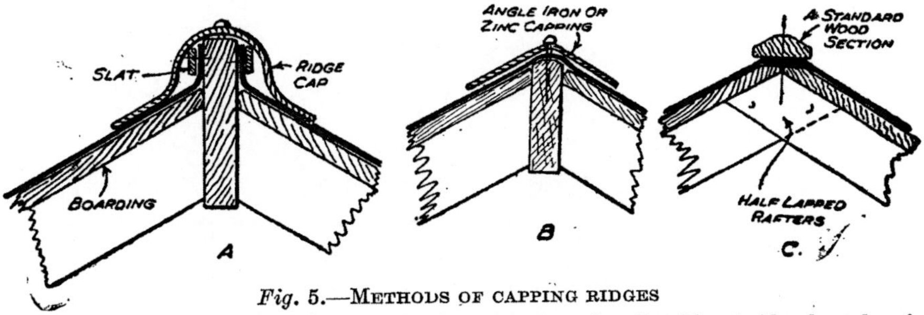

Fig. 5.—METHODS OF CAPPING RIDGES

A, with raised ridge board. B, with flush ridge board. C, without ridge board, using wood capping.

be where it is simply nailed on and left unembellished in any way at its edges, it can nevertheless be used in a manner which renders it both less distasteful and incongruous among other more traditional buildings.

Corrugated-iron roofs are very hot in summer and cold in winter. By combining corrugated iron with a layer of wood boarding (with or without a layer or roofing felt) or insulating board, a very much better roof and a more habitable interior can be obtained. If the ridge is capped with, say, a half-round tile set in cement, the iron painted a red or brown colour, and the sheeting used in one length as is possible in a small building, quite a respectable finish may be obtained, especially if the gables and eaves are provided with a barge board and a fascia board with guttering as illustrated in the accompanying Figs. 8, 9, and 10.

Further, the roof will be quite weather-tight, and the dead air space between the wood and the iron will form a fairly good heat insulation.

Laying Corrugated Sheets

Fig. 8 shows the line of roofing, illustrating on the left the treatment at the gable end, and on the right the joint between sheetings supported

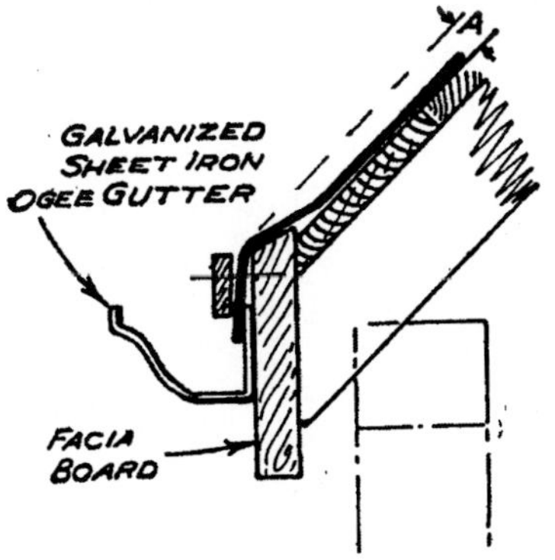

Fig. 6.—FINISHING OFF AT EAVES WITH GUTTERING

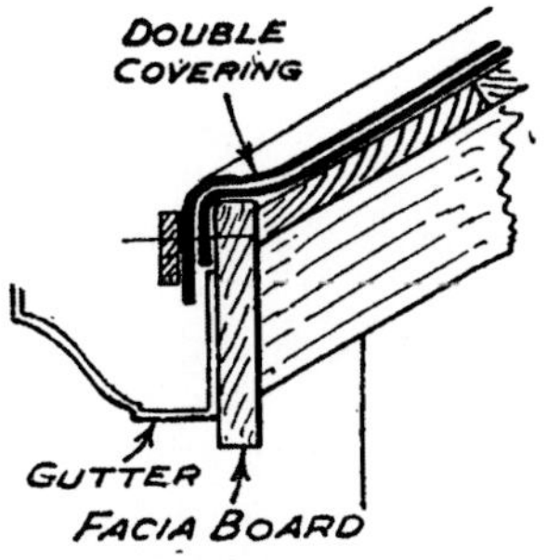

Fig. 7.—ANOTHER METHOD OF FINISHING AT EAVES WITH GUTTERING

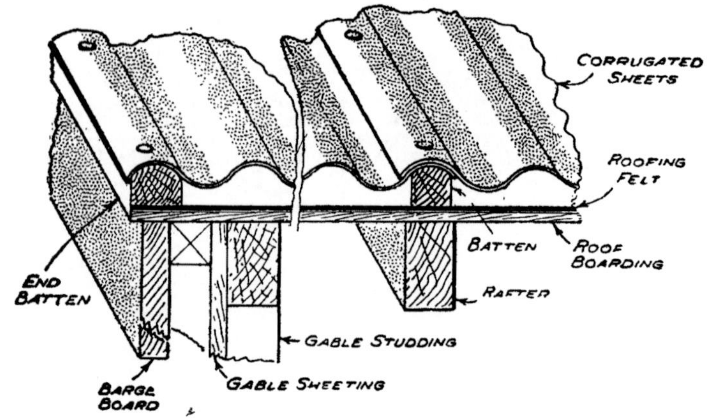

Fig. 8.—ROOFING WITH CORRUGATED-IRON SHEETS

This shows section of roof with boarding and roofing felt under corrugated sheet covering.

by a continuous batten planed to fit in the hollow of the corrugations. Sheets up to 12 ft. long are obtainable, although 10 ft. is a more usual maximum length. This means that a span of about 18 or 19 ft. can be bridged by a single sheet on each side of the roof.

Treatment of Ridges and Eaves

At ridges and eaves the details can be arranged as shown in sketch (Fig. 9). At the ridges firring may be run along the ridge board as indicated to support the sheet, and the eaves fascia board may be raised up to support the other end of the sheeting. An alternative arrangement is indicated in Fig. 10, where the design demands an overhanging eaves and a more ornate guttering.

With single sheets only screw roofing nails will be required.

Corrugated iron is 2 ft. 3 in. wide and fits on battens placed at 2-ft. centres. Details of corrugated asbestos sheeting are given in Chapter VII. Various sizes of asbestos sheet are available, and if the above method of construction is adopted the battens and ridge boards will have to be adapted to the size of sheet.

FIXING RUBEROID ROOFING

Ruberoid is a trade name for a roofing material made in several forms. The standard stuff is sent out in rolls 3 ft. wide. Besides the ordinary grey roofing it can be obtained in a permanent red or green colour, and in two finishes, viz. smooth and slate surfaces. Where a more decorative form is required, Ruberoid is made in strips punched out so that the finished effect is that of slates of either a red or green colour.

It is advisable that Ruberoid should have a sound foundation. A grooved and tongued under-boarding is always to be preferred to sawn boards, as it lies flatter, in spite of any vee groove in its edges. For

Ruberoid, ¾-in. tongued and grooved floor-boarding is recommended, otherwise 1-in. close-butted sawn boarding should be used.

In either case, the board should be well nailed to the rafters or other supports, with the heads of all brads punched below the surface.

Where rough (sawn) boards are used, any upstanding edges should be planed off so that a relatively smooth, even surface is obtained. The supports for the board should be sufficiently stiff for the particular job in hand.

All arrises over which the covering has to be folded or bent should be chamfered off in a rounded form to prevent any tendency to cut through the material.

Preliminary Treatment

The material requires anything up to fourteen days' weathering before use. It should be laid in an open space, after being cut, so that it may expand, and, naturally, this treatment is more rapidly accomplished in the summer than in winter. The material is best laid in warm weather, and in any case periods of rain and snow should be avoided. In cold and frosty weather it may be necessary to heat the roof covering with a blow-lamp flame, used by rapidly passing it over the surface of the sheet at the required spot before attempting to bend or fold the material.

The outside of the roll of roofing should always be laid downwards, with that face next to the boarding.

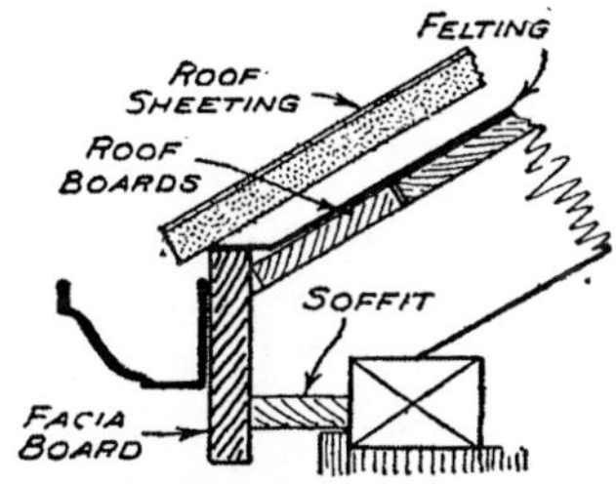

Fig. 10 (*above*).—ALTERNATIVE ARRANGEMENT OF CORRUGATED ROOF WITH OVERHANGING EAVES

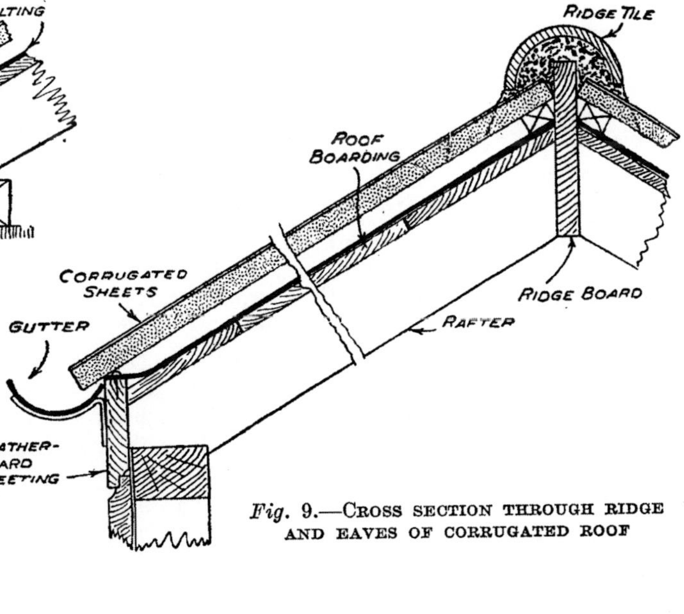

Fig. 9.—CROSS SECTION THROUGH RIDGE AND EAVES OF CORRUGATED ROOF

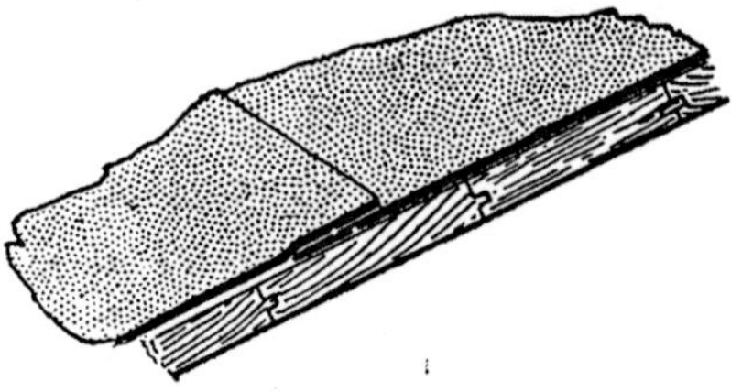

Incorrect

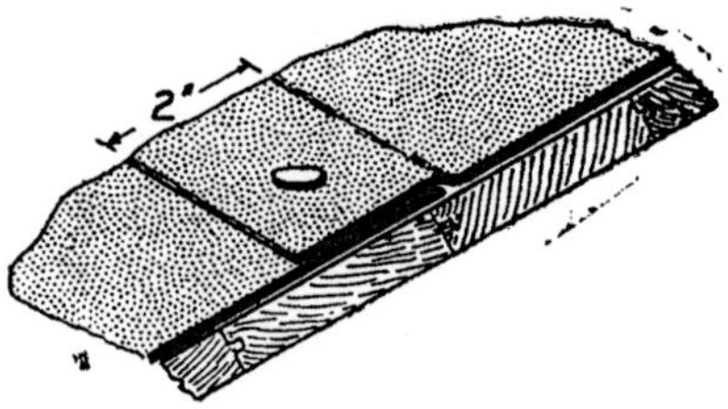

Correct

Figs. 11 *and* 12.—HOW TO LAY RUBEROID ROOFING HORIZONTALLY

Horizontal joints on a pitch roof or on a fall in a flat roof should be made as at Fig. 12, not as shown in Fig. 11. The lap should be 2 in. minimum.

Horizontal Joints

The proper way to arrange the horizontal lap joints is shown in Fig. 12.

After the lap joint is prepared and cemented, the nails—$\frac{3}{4}$-in. galvanised clouts are used in the ordinary course—may be driven through the joint. The extent of the lap should not be less than 2 in., and the pitch of the nails should be an equal amount. The seam of the two sheets is made water-tight with a solvent compound applied by a brush. This material must not be heated, although it is not quite so adhesive or free-flowing in very cold weather. It may be kept warm but must not be put over a fire like pitch. The compound does not adhere to a wet surface.

After rain or melting snow the roof must be allowed to dry off before proceeding with the work. In addition, work in completing the joints must be held over if the material has not had sufficient weathering time before laying is commenced.

Roof of Flat Pitch

On ordinary roofs of the flatter pitches it is advisable to fix Ruberoid with the joints running from the ridges to the eaves, i.e. with downwards joints (Fig. 13), unless other considerations predominate. The slate-surfaced material should, for instance, always be laid from verge to verge, viz. with the joints parallel to the eaves of the roof, except where the under-boarding is thin or springy. In this case, the arrangement already referred to (Fig. 13) is better, the joints of the covering crossing those of the boards at right angles.

No rolls or lifting strips are required with Ruberoid, as the cementing renders "down-roof" joints and nails quite weather-tight. There is no need therefore to run the roofing material over the ridge. The latter can be almost flush with the surface of the boarding and have any raw edges planed off, the ridge cap being formed by bending an 8-in. or 9-in. strip of the roofing material to the angle required, cementing and clout-nailing the edges down, as shown in Fig. 14.

Cementing the Joints

Each length of roofing material should lap 2 in. and, in applying the cement, first remove any powder from the edge or surface to be coated and put on the compound liberally, working it in with the brush, as shown in Fig. 13. Nails should be at least 1 in. from the edge of the sheet and, when laid, coat the nail heads with the cement. A good finish is obtained by painting a band 2 in. or $2\frac{1}{2}$ in. wide down the joint, using chalk lines to guide the brush in a neat straight course. For the slate-surfaced material, painting the nail heads is not recommended.

Gutters and Valleys

In covering a roof having gutters or valleys these parts should be fitted and finished before the run of the roof is commenced. For gutters, two thicknesses of Ruberoid, bedded together with the mastic compound, are recommended. The strips should be laid in as long a length as possible to reduce the number of joints to the minimum. It is important not to use any clout nails in the bottom of the gutter ; not in the upper layer at any rate. The only nailing that is necessary is at the joints. With this particular material, the carpenter need not provide any drip steps in the fall of the gutter such as are essential in providing the supporting woodwork for zinc or lead. Therefore, in calculating and marking off the fall of the gutter, there is no need to allow extra height for these drips. A straight fall from end to end of 2 in. in 10 ft. is all that is necessary.

Flashing to Brick Walls

For flashing a more or less horizontal joint where a brick wall abuts on to a flat roof or to a gutter, the brick joint not more than two courses up must be raked out to a depth of not less than 1 in. A strip of roofing

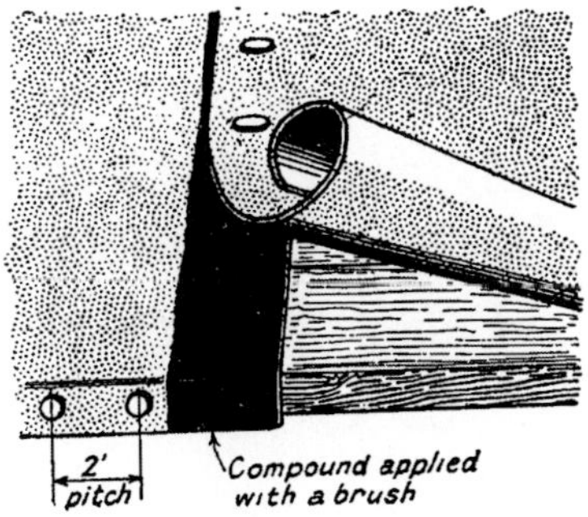

Fig. 13.—How Ruberoid Down-roof seams are nailed and cemented

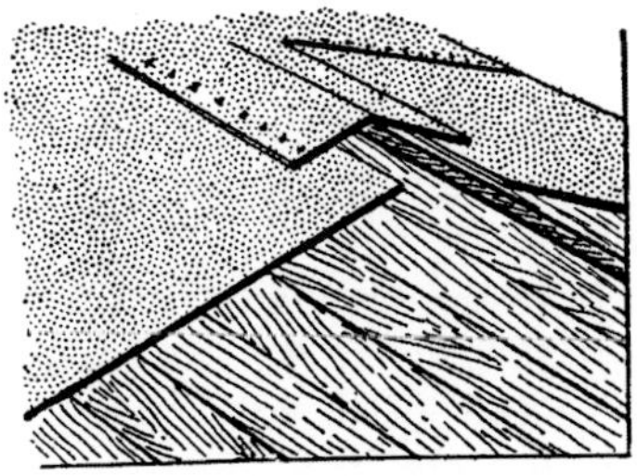

Fig. 14.—Ridge capping

This may be formed out of an 8- or 9-in. strip of the roofing material. Joint nailed and cemented.

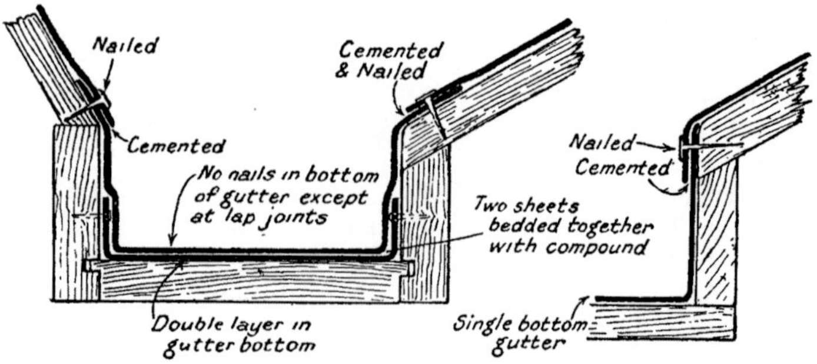

Fig. 15.—Gutter between saw tooth roofs with double bottom of Ruberoid

Fig. 16.—Alternative arrangement with single layer of Ruberoid

material must then be cut off, in as long a length as possible and of such a width as will allow a ¾-in. bend into the wall and a 4-in. turn out on to the flat. If the weather is cold and frosty, the strip of Ruberoid may be warmed by a flash of a blow-lamp, as previously referred to.

The main roofing, or gutter bottom, as the case may be, should turn up the wall 2 in. or 3 in. at least, and be spiked into the nearest joint above the surface of the gutter or roof. The flashing strip, being fitted, can then be secured by wood or lead wedges, as shown in Fig. 19, at the "turn in" and with mastic compound and galvanised clouts along the lower flanged outer edge, as indicated. The job may then be passed back to the bricklayer to make good the top joint by Portland cement pointing. The flat joint and nail-heads should be painted over with the compound in the usual way to ensure water-tightness.

An internal angle is dealt with in the same manner, the flashings at the corner being cut out and laid as shown in Fig. 22. The left-hand piece of flashing, that cut at an angle as at A, overlaps the other. Wood or lead wedge plugs are used in the raked-out joint in the wall, as in the fixing of a plain run of flashing, clout nails being driven through the overlaps, as shown. A plumber's lead-dressing tool, or its equivalent, will be found useful in forming internal angles in the Ruberoid covering. An external angle is dealt with in a similar

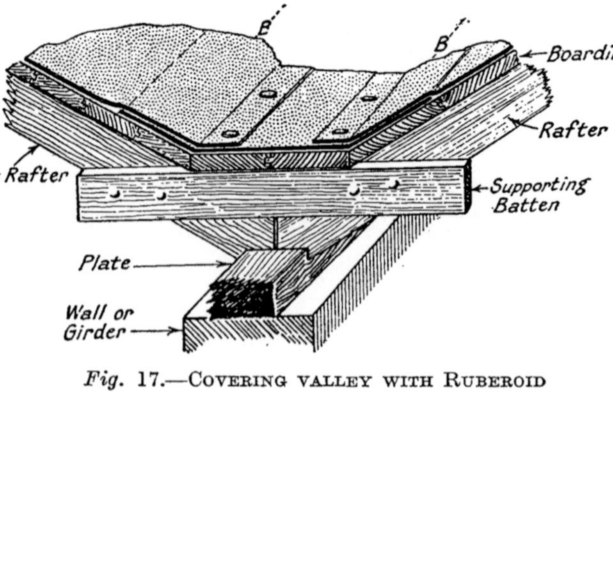

Fig. 17.—Covering valley with Ruberoid

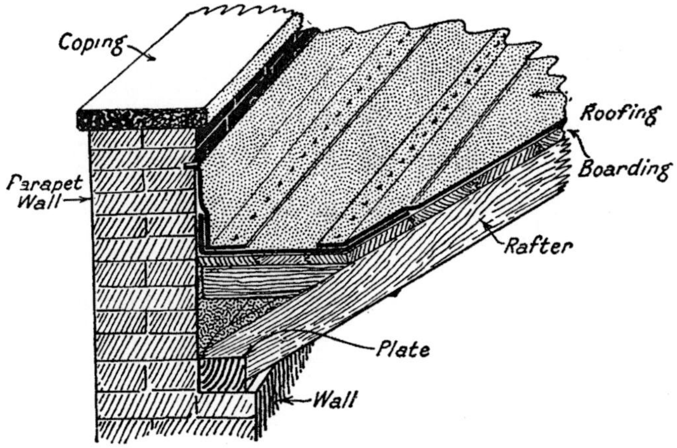

Fig. 18.—Flashing of Ruberoid to brick wall

manner, the details, Fig. 23, showing the cutting and the fixing.

Stepped Flashing

Stepped flashing can be arranged in Ruberoid just as easily as in lead or zinc. The flashing strips are cut out in triangular form, as illustrated in Figs. 20 and 21, and are fixed in the same way. At a ridge the piece is of a duplex form. The dimensions will depend on the pitch of the roof. If the upright wall is of wood, then the flashing may be clout-nailed instead of being turned into a brick joint.

Ridges, Eaves, and Verges

At ridges, the detail illustrated in Fig. 14 may be followed, the strip forming the capping being the last part to be fixed. Clout nails may be relied on to fix this roofing material on the vertical edges of eaves and verges (gable ends). The mechanical strength of Ruberoid is much superior to ordinary bitumen roofing felt, and as the heads of the nails may be painted over with the mastic compound after they are driven, there is no danger of wet-weather conditions rotting the fixings. At the same time it is important that the nails should be driven into wood of a thickness that will satisfactorily support them. Therefore the use of a thicker board, and an essentially sound one, at any edge formed in the roof around which the roofing has to be bent and nailed, are points which should be borne in mind. It is also obvious that nailing in or too near to a joint in the under-boarding should be avoided, and in any case, end grain is none too good, even where the nails are supporting roofing material.

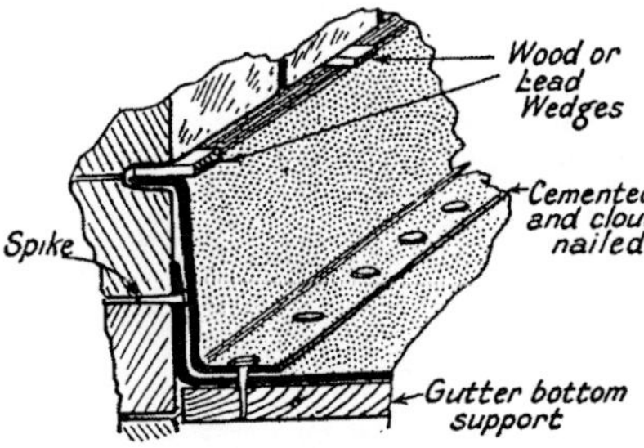

Fig. 19.—Making joint with wall

The flashing strip is turned into the open joint, wedged, and the apron nailed and cemented down to the roof surface.

Double-layer Roofs

Where a double layer of material on a pitched roof is required, the

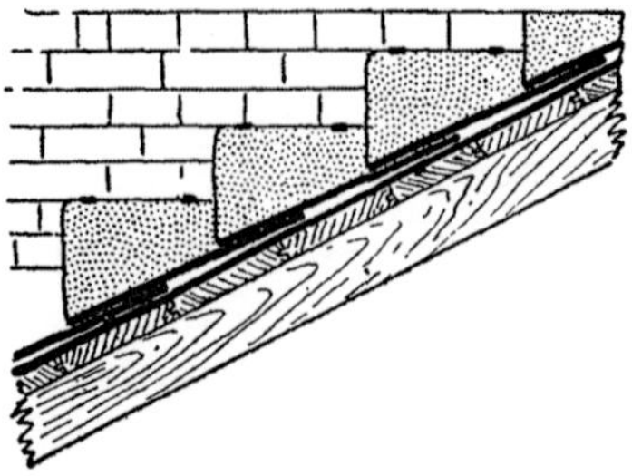

Fig. 20.—STEPPED FLASHINGS FOR AN INCLINED ROOF FIXED IN POSITION

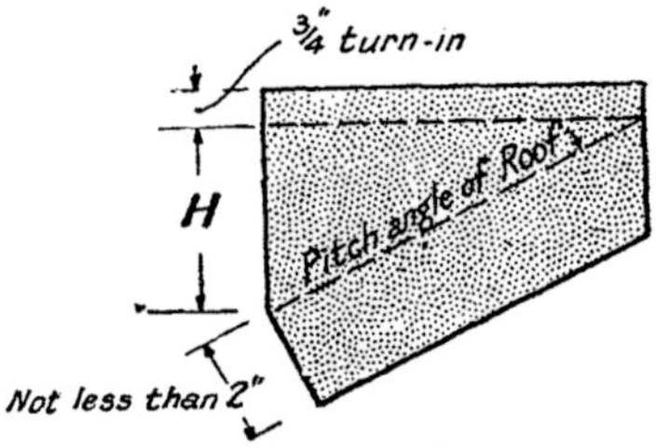

Fig. 21.—STEPPED FLASHINGS AS CUT

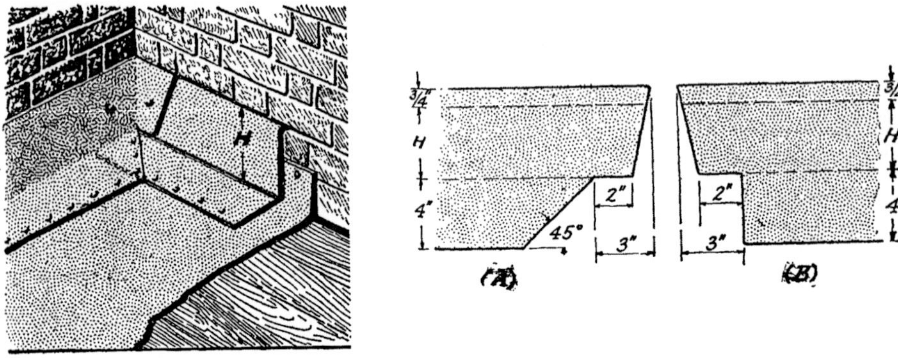

Fig. 22.—RUBEROID FLASHINGS TO AN INTERNAL ANGLE OF WALL

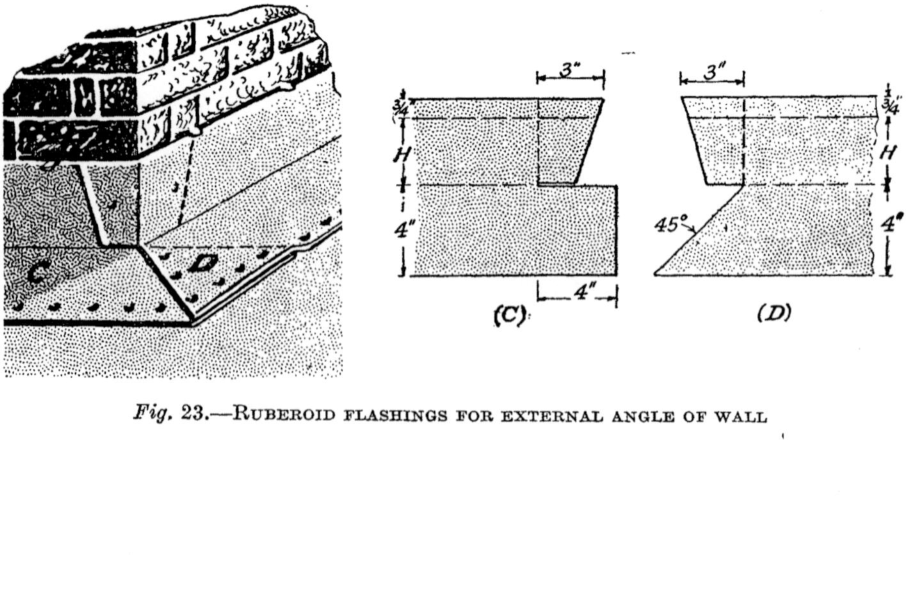

Fig. 23.—RUBEROID FLASHINGS FOR EXTERNAL ANGLE OF WALL

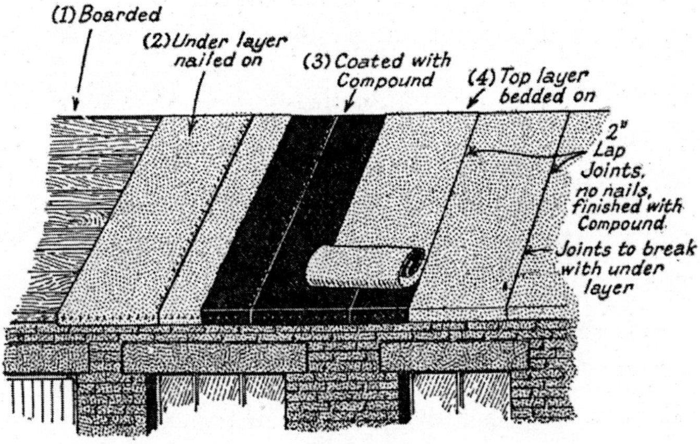

Fig. 24.—Laying roof with double thickness of Ruberoid

system often employed is illustrated in Fig. 24. The under layer is fixed with "down-roof" joints with nails only, and similarly secured at eaves and ridges. The second stage is to coat the whole surface with mastic compound and, as these processes extend, to put on the outer layer of covering. The usual 2-in. lap joint should be provided in fixing adjacent outer sheets, but all nailing may be omitted, weather-tightness being ensured by painting all edge joints with compound. Care must be taken to press out all bubbles of air in laying this outer covering. Ruberoid is absolutely air-tight if properly secured, but if the air is not squeezed out, it will remain trapped, causing expansion or contraction according to temperature.

Covering Open Roofs

Where the material is required to cover roofs which are open underneath and in which the under-boards are not of the grooved and tongued form, it may be necessary to support the sheets by a strip of Ruberoid, 1½ in. wide, nailed and cemented down the centre of each sheet. This, of course, only applies to a roof covered with "down-roof" joints.

The Finish at Eaves

A feather-edge board, or its equivalent, can be used to ensure a sufficient thickness of wood in clout-nailing the turned-over edge of the Ruberoid at the eaves. The edge of the Ruberoid should overlap the back flange of metal gutters.

ROOFING TABLES

Compiled by E. W. BARBER

CERTAIN rules for finding sizes of roof members have already been given on p. 22. Calculations regarding strength and safe loads will be found in Chapter IV, which begins on p. 43. In the same chapter will also be found a method of obtaining the correct bevels for common rafters, hip rafters, jack rafters, roof boarding, and purlins.

In actual practice it is necessary for a carpenter to have a reliable means of ascertaining the lengths of common rafters, jack rafters, and backed hips to suit the particular roof truss upon which he is working. In the hands of anyone who is thoroughly familiar with its use, the steel square is a most useful tool for this purpose.

Readers who wish to make themselves acquainted with the method of using the steel square are advised to obtain a specialised book on the subject, e.g. *The Steel Square,* by Noel D. Green (George Newnes, Ltd.).

As an alternative, we give in the following pages a series of tables which may be used to obtain the same results.

HOW TO USE THE ROOFING TABLES

By the use of the roofing tables, roof members for a pitched roof (of any one of the eight pitches given) may be measured to the correct lengths and bevels found so that the members may be cut without the aid of the steel square or any other patent device, regardless of span. In compiling these tables, the span of the building has been taken as the overall dimension including the external walls.

The first table, which occupies pp. 128 and 129, gives the lengths of common rafter for spans ranging from 10 ft. to 30 ft. 11 in.

Table II, which occupies pp. 130 and 131, shows the correct lengths of backed hips for roofs of different pitches having any span between 10 ft. and 30 ft. 11 in.

Bevels for roof members can be obtained from the table on p. 132, whilst general data will be found on p. 133.

The following examples are worked out for a hip roof with a span of 22 ft. 4 in., the pitch to be $\frac{3}{8}$.

Length of Common Rafter

Refer to Table I, headed "Length of Common Rafter." On p. 129 find the column headed 22 (feet) and read downwards to the horizontal line corresponding to the required pitch (in this case $\frac{3}{8}$). This gives a length of common rafter 13 ft. 9 in. for a roof having a span of 22 ft. and a pitch of $\frac{3}{8}$. Read along the top of left-hand table to 22 ft., but as the span is 22 ft. 4 in. it is necessary to refer to the right hand of table, which deals with the inches. Read along the top of table to 4 in. and read down to required $\frac{3}{8}$ pitch. This reads $2\frac{1}{2}$ in. These two results are added together, thus 13 ft. 9 in. + $2\frac{1}{2}$ in. = 13 ft. $11\frac{1}{2}$ in., this being the length of common rafter for a roof with a span of 22 ft. with $\frac{3}{8}$ pitch.

Length of Backed Hips

Refer to Table II, headed "Length of Backed Hips," and repeat as for common rafter, the results reading 17 ft. $7\frac{5}{16}$ in. + $3\frac{3}{16}$ in. = 17 ft. $10\frac{1}{2}$ in. Valleys the same.

Length of Jack Rafters

Refer to Table III, last column, which is headed "Diminishing Length of Jack Rafters." Read downwards to the line corresponding to $\frac{3}{8}$ pitch at left of table, which is 1 ft. $3\frac{5}{16}$ in. This is marked on common rafter, starting from the long point of plumb cut and repeated along length of rafter. This gives the different lengths and number of jack rafters down the hip at 14-in. centres.

Length of Jack Rafters

The following example is worked out for a roof with a span of 16 ft., the pitch to be 45° Reference to Table I, headed "Length of Common Rafter," will show that the common rafter in this case will be 11 ft. $3\frac{3}{4}$ in.

Now refer to Table III, headed "Roof Members. Cuts in 8 Pitches." Read down column 1, which gives the slope of roof in degrees, and 45° ($\frac{1}{2}$ pitch of roof) will be found. Reading across the table to the final column, the diminishing length of jack rafter is given as 1 ft. $5\frac{1}{4}$ in. The jack rafters are centred down the hip at 14 in., so that it is now possible to determine both the length and number of jack rafters required.

The common rafter is 11 ft. $3\frac{3}{4}$ in., and the diminishing length of the jack rafter is 1 ft. $5\frac{1}{4}$ in. By simple subtraction we find that the length of the jack rafter (A) is 9 ft. $10\frac{1}{2}$ in. By repeating the process—9 ft. $10\frac{1}{2}$ in. less 1 ft. $5\frac{1}{4}$ in.—we get the length of rafter (B) as 8 ft. $5\frac{1}{4}$ in., and so on.

In actual practice the diminishing length of the jack rafter is marked on the common rafter, starting from the long point of plumb cut and repeated along the length of the rafter.

The Bevels of Roof Members

Refer to Table III, headed "Roof Members. Cuts in 8 Pitches." At the top of this section are the names of the members, and as each one is required read downwards to the required pitch on the left of this section. See below :

36° 52′	9	$\frac{3}{8}$	37°	53°	28°	62°	41°	65°	38°	58°	51°	1′ $3\frac{5}{16}$″

The jack rafter plumb cut and bevel cut is the same as the plumb cut and bevel cut of common rafter.

TABLE I.—LENGTH OF COMMON RAFTER

Feet ←—— *Span of the Building* ——→ *Inches*

Slope of Roof in Degrees		Rise of Roof in 12 in. (½ Span)	Pitch of Roof	10		11		12		13		14	
°	′	*in.*		*ft.*	*in.*	*ft.*	*in.*	*ft.*	*in.*	*ft.*	*in.*	*ft.*	*in.*
14	2	3	1/8	5	1 7/8	5	8	6	2 1/4	6	8 7/16	7	2 9/16
18	26	4	1/6	5	3 1/4	5	9 9/16	6	3 7/8	6	10 3/16	7	4 9/16
26	34	6	1/4	5	7 1/16	6	1 13/16	6	8 1/2	7	3 3/16	7	9 15/16
33	41	8	1/3	6	0 1/8	6	7 5/16	7	2 9/16	7	9 3/4	8	4 15/16
36	52	9	3/8	6	3	6	10 1/2	7	6	8	1 1/2	8	9
45	0	12	1/2	7	0 7/8	7	9 5/16	8	5 13/16	9	2 5/16	9	10 7/8
53	8	16	2/3	8	4	9	2	10	0	10	10	11	8
56	19	18	3/4	9	0 1/8	9	11	10	9 13/16	11	8 5/8	12	7 7/16

Slope of Roof in Degrees		Rise of Roof in 12 in. (½ Span)	Pitch of Roof	1	2	3	4	5	6	7	8	9	10	11
°	′	*in.*		*in.*	*in.*	*in.*	*in.*	*in.*	*in.*	*in.*	*in.*	*in.*	*in.*	*in.*
14	2	3	1/8	1/2	1	1 9/16	2 1/16	2 9/16	3 1/16	3 9/16	4 1/8	4 5/8	5 1/8	5 11/16
18	26	4	1/6	1/2	1 1/16	1 9/16	2 1/8	2 5/8	3 3/16	3 11/16	4 3/16	4 3/4	5 1/4	5 13/16
26	34	6	1/4	9/16	1 1/8	1 11/16	2 1/4	2 13/16	3 3/8	3 15/16	4 1/2	5	5 9/16	6 1/8
33	41	8	1/3	5/8	1 3/16	1 13/16	2 3/8	3	3 5/8	4 3/16	4 13/16	5 3/8	6	6 5/8
36	52	9	3/8	5/8	1 1/4	1 7/8	2 1/2	3 1/8	3 3/4	4 3/8	5	5 5/8	6 1/4	6 7/8
45	0	12	1/2	11/16	1 7/16	2 1/8	2 13/16	3 9/16	4 1/4	4 15/16	5 11/16	6 3/8	7 1/16	7 3/4
53	8	16	2/3	13/16	1 11/16	2 1/2	3 5/16	4 3/16	5	5 13/16	6 11/16	7 1/2	8 5/16	9 3/16
56	19	18	3/4	7/8	1 13/16	2 11/16	3 5/8	4 1/2	5 7/16	6 5/16	7 3/16	8 1/8	9	9 15/16

Slope of Roof in Degrees		Rise of Roof in 12 in. (½ Span)	Pitch of Roof	15		16		17		18		19	
°	′	*in.*		*ft.*	*in.*	*ft.*	*in.*	*ft.*	*in.*	*ft.*	*in.*	*ft.*	*in.*
14	2	3	1/8	7	8 3/4	8	2 15/16	8	9 1/8	9	3 5/16	9	9 1/2
18	26	4	1/6	7	10 7/8	8	5 3/16	8	11 1/2	9	5 13/16	10	0 3/16
26	34	6	1/4	8	4 5/8	8	11 5/16	9	6 1/4	10	0 3/4	10	7 7/16
33	41	8	1/3	9	0 3/16	9	7 3/8	10	2 9/16	10	9 13/16	11	5
36	52	9	3/8	9	4 1/2	10	0	10	7 1/2	11	3	11	10 1/2
45	0	12	1/2	10	7 5/16	11	3 3/4	12	0 1/4	12	8 3/4	13	5 3/16
53	8	16	2/3	12	6	13	4	14	2	15	0	15	10
56	19	18	3/4	13	6 1/4	14	5 1/16	15	3 7/8	16	2 11/16	17	1 1/2

Slope of Roof in Degrees		Rise of Roof in 12 in. (½ Span)	Pitch of Roof	1	2	3	4	5	6	7	8	9	10	11
°	′	*in.*		*in.*	*in.*	*in.*	*in.*	*in.*	*in.*	*in.*	*in.*	*in.*	*in.*	*in.*
14	2	3	1/8	1/2	1	1 9/16	2 1/16	2 9/16	3 1/16	3 9/16	4 1/8	4 5/8	5 1/8	5 11/16
18	26	4	1/6	1/2	1 1/16	1 9/16	2 1/8	2 5/8	3 3/16	3 11/16	4 3/16	4 3/4	5 1/4	5 13/16
26	34	6	1/4	9/16	1 1/8	1 11/16	2 1/4	2 13/16	3 3/8	3 15/16	4 1/2	5	5 9/16	6 1/8
33	41	8	1/3	5/8	1 3/16	1 13/16	2 3/8	3	3 5/8	4 3/16	4 13/16	5 3/8	6	6 5/8
36	52	9	3/8	5/8	1 1/4	1 7/8	2 1/2	3 1/8	3 3/4	4 3/8	5	5 5/8	6 1/4	6 7/8
45	0	12	1/2	11/16	1 7/16	2 1/8	2 13/16	3 9/16	4 1/4	4 15/16	5 11/16	6 3/8	7 1/16	7 3/4
53	8	16	2/3	13/16	1 11/16	2 1/2	3 5/16	4 3/16	5	5 13/16	6 11/16	7 1/2	8 5/16	9 3/16
56	19	18	3/4	7/8	1 13/16	2 11/16	3 5/8	4 1/2	5 7/16	6 5/16	7 3/16	8 1/8	9	9 15/16

If span is in feet and inches, find result of feet span in left-hand table. Then find result of the inches of span in right-hand table and add two results together.

TABLE I (*contd.*)—LENGTH OF COMMON RAFTER

Feet ←——— *Span of the Building* ———→ *Inches*

Slope of Roof in Degrees (° ′)	Rise of Roof in 12 in. (1/2 Span) in.	Pitch of Roof	20 ft.	20 in.	21 ft.	21 in.	22 ft.	22 in.	23 ft.	23 in.	24 ft.	24 in.
14 2	3	1/8	10	3 11/16	10	9 7/8	11	4 1/16	11	10 1/4	12	4 7/16
18 26	4	1/6	10	6 1/2	11	0 13/16	11	7 1/8	12	1 7/16	12	7 13/16
26 34	6	1/4	11	2 3/16	11	8 7/8	12	3 9/16	12	10 5/16	13	5
33 41	8	1/3	12	0 3/16	12	7 7/16	13	2 1/2	13	9 7/8	14	5 1/16
36 52	9	3/8	12	6	13	1 1/2	13	9	14	4 1/2	15	0 1/2
45 0	12	1/2	14	1 3/4	14	10 3/16	15	6 11/16	16	3 3/16	16	11 5/8
53 8	16	2/3	16	8	17	6	18	4	19	2	20	0
56 19	18	3/4	18	0 5/16	18	11 1/8	19	9 15/16	20	8 13/16	21	7 9/16

Slope of Roof in Degrees (° ′)	Rise of Roof in 12 in. (1/2 Span) in.	Pitch of Roof	1 in.	2 in.	3 in.	4 in.	5 in.	6 in.	7 in.	8 in.	9 in.	10 in.	11 in.
14 2	3	1/8	1/2	1	1 9/16	2 1/16	2 9/16	3 1/16	3 9/16	4 1/8	4 5/8	5 1/8	5 11/16
18 26	4	1/6	1/2	1 1/16	1 9/16	2 1/8	2 5/8	3 3/16	3 11/16	4 3/16	4 3/4	5 1/4	5 13/16
26 34	6	1/4	9/16	1 1/8	1 11/16	2 1/4	2 13/16	3 3/8	3 15/16	4 1/2	5	5 9/16	6 1/8
33 41	8	1/3	5/8	1 3/16	1 13/16	2 3/8	3	3 5/8	4 3/16	4 13/16	5 3/8	6	6 5/8
36 52	9	3/8	5/8	1 1/4	1 7/8	2 1/2	3 1/8	3 3/4	4 3/8	5	5 5/8	6 1/4	6 7/8
45 0	12	1/2	11/16	1 7/16	2 1/8	2 13/16	3 9/16	4 1/4	4 15/16	5 11/16	6 3/8	7 1/16	7 3/4
53 8	16	2/3	13/16	1 11/16	2 1/2	3 5/16	4 3/16	5	5 13/16	6 11/16	7 1/2	8 5/16	9 3/16
56 19	18	3/4	7/8	1 13/16	2 11/16	3 5/8	4 1/2	5 7/16	6 5/16	7 3/16	8 1/8	9	9 15/16

Slope of Roof in Degrees (° ′)	Rise of Roof in 12 in. (1/2 Span) in.	Pitch of Roof	25 ft.	25 in.	26 ft.	26 in.	27 ft.	27 in.	28 ft.	28 in.	29 ft.	29 in.	30 ft.	30 in.
14 2	3	1/8	12	10 5/8	13	4 13/16	13	11	14	5 3/16	14	11 3/8	15	5 9/16
18 26	4	1/6	13	2 1/8	13	8 7/16	14	2 3/4	14	9 1/16	15	3 3/8	15	9 3/4
26 34	6	1/4	13	11 11/16	14	6 7/16	15	1 1/8	15	7 11/16	16	2 7/16	16	9 1/8
33 41	8	1/3	15	0 1/4	15	7 1/2	16	2 11/16	16	9 15/16	17	5 1/8	18	0 15/16
36 52	9	3/8	15	7 1/2	16	3	16	10 1/2	17	6	18	1 1/2	18	9
45 0	12	1/2	17	8 1/8	18	4 5/8	19	1 1/8	19	9 9/16	20	6 1/16	21	2 9/16
53 8	16	2/3	20	10	21	8	22	6	23	4	24	2	25	0
56 19	18	3/4	22	6 7/16	23	5 1/4	24	4 1/16	25	2 7/8	26	1 11/16	27	0 1/2

Slope of Roof in Degrees (° ′)	Rise of Roof in 12 in. (1/2 Span) in.	Pitch of Roof	1 in.	2 in.	3 in.	4 in.	5 in.	6 in.	7 in.	8 in.	9 in.	10 in.	11 in.
14 2	3	1/8	1/2	1	1 9/16	2 1/16	2 9/16	3 1/16	3 9/16	4 1/8	4 5/8	5 1/8	5 11/16
18 26	4	1/6	1/2	1 1/16	1 9/16	2 1/8	2 5/8	3 3/16	3 11/16	4 3/16	4 3/4	5 1/4	5 13/16
26 34	6	1/4	9/16	1 1/8	1 11/16	2 1/4	2 13/16	3 3/8	3 15/16	4 1/2	5	5 9/16	6 1/8
33 41	8	1/3	5/8	1 3/16	1 13/16	2 3/8	3	3 5/8	4 3/16	4 13/16	5 3/8	6	6 5/8
36 52	9	3/8	5/8	1 1/4	1 7/8	2 1/2	3 1/8	3 3/4	4 3/8	5	5 5/8	6 1/4	6 7/8
45 0	12	1/2	11/16	1 7/16	2 1/8	2 13/16	3 9/16	4 1/4	4 15/16	5 11/16	6 3/8	7 1/16	7 3/4
53 8	16	2/3	13/16	1 11/16	2 1/2	3 5/16	4 3/16	5	5 13/16	6 11/16	7 1/2	8 5/16	9 3/16
56 19	18	3/4	7/8	1 13/16	2 11/16	3 5/8	4 1/2	5 7/16	6 5/16	7 3/16	8 1/8	9	9 15/16

If span is in feet and inches, find result of feet span in left-hand table. Then find result of the inches of span in right-hand table and add two results together.

TABLE II.—LENGTH OF BACKED HIPS

Feet ←——— *Span of the Building* ———→ *Inches*

Slope of Roof in Degrees °	′	Rise of Roof in 12 in. (½ Span) in.	Pitch of Roof	10 ft.	10 in.	11 ft.	11 in.	12 ft.	12 in.	13 ft.	13 in.	14 ft.	14 in.
14	2	3	1/8	7	2 3/16	7	10 13/16	8	7 3/8	9	4	10	0 5/8
18	26	4	1/6	7	3 3/16	7	11 7/8	8	8 5/8	9	5 5/16	10	2 1/16
26	34	6	1/4	7	6	8	3	9	0	9	9	10	6
33	41	8	1/3	7	9 13/16	8	7 3/16	9	4 9/16	10	1 15/16	10	11 5/16
36	52	9	3/8	8	0 1/16	8	9 5/8	9	7 1/4	10	4 7/8	11	2 7/16
45	0	12	1/2	8	7 15/16	9	6 5/16	10	4 11/16	11	3 1/8	12	1 1/2
53	8	16	2/3	9	8 5/8	10	8 1/4	11	7 15/16	12	7 5/8	13	7 1/4
56	19	18	3/4	10	3 11/16	11	2 7/8	12	4 7/16	13	4 13/16	14	5 3/16

Slope of Roof in Degrees °	′	Rise of Roof in 12 in. (½ Span) in.	Pitch of Roof	1 in.	2 in.	3 in.	4 in.	5 in.	6 in.	7 in.	8 in.	9 in.	10 in.	11 in.
14	2	3	1/8	11/16	1 7/16	2 1/8	2 7/8	3 9/16	4 5/16	5	5 3/4	6 7/16	7 3/16	7 7/8
18	26	4	1/6	3/4	1 7/16	2 3/16	2 7/8	3 5/8	4 3/8	5 1/16	5 13/16	6 9/16	7 1/4	8
26	34	6	1/4	3/4	1 1/2	2 1/4	3	3 3/4	4 1/2	5 1/4	6	6 3/4	7 1/2	8 1/4
33	41	8	1/3	13/16	1 9/16	2 3/8	3 1/8	3 15/16	4 11/16	5 1/2	6 1/4	7 1/16	7 13/16	8 5/8
36	52	9	3/8	13/16	1 9/16	2 3/8	3 3/16	4	4 13/16	5 5/8	6 7/16	7 3/16	8	8 13/16
45	0	12	1/2	7/8	1 3/4	2 5/8	3 1/2	4 5/16	5 3/16	6 1/16	6 15/16	7 13/16	8 11/16	9 9/16
53	8	16	2/3	15/16	1 15/16	2 15/16	3 7/8	4 15/16	5 13/16	6 13/16	7 3/4	8 3/4	9 11/16	10 11/16
56	19	18	3/4	1	2 1/16	3 1/16	4 1/8	5 1/8	6 3/16	7 3/16	8 1/4	9 1/4	10 5/16	11 5/16

Slope of Roof in Degrees °	′	Rise of Roof in 12 in. (½ Span) in.	Pitch of Roof	15 ft.	15 in.	16 ft.	16 in.	17 ft.	17 in.	18 ft.	18 in.	19 ft.	19 in.
14	2	3	1/8	10	9 1/4	11	5 7/8	12	2 1/2	12	11 1/8	13	7 3/4
18	26	4	1/6	10	10 11/16	11	7 1/2	12	4 3/16	13	0 15/16	13	9 5/8
26	34	6	1/4	11	3	12	0	12	9	13	6	14	3
33	41	8	1/3	11	8 11/16	12	6 1/16	13	3 1/2	14	0 7/8	14	10 1/4
36	52	9	3/8	12	0 1/16	12	9 11/16	13	7 1/4	14	4 7/8	15	2 1/2
45	0	12	1/2	12	11 7/8	13	10 1/4	14	8 11/16	15	7 1/16	16	5 7/16
53	8	16	2/3	14	6 15/16	15	6 5/8	16	6 1/4	17	5 15/16	18	5 9/16
56	19	18	3/4	15	5 9/16	16	5 15/16	17	6 5/16	18	6 5/8	19	7

Slope of Roof in Degrees °	′	Rise of Roof in 12 in. (½ Span) in.	Pitch of Roof	1 in.	2 in.	3 in.	4 in.	5 in.	6 in.	7 in.	8 in.	9 in.	10 in.	11 in.
14	2	3	1/8	11/16	1 7/16	2 1/8	2 7/8	3 9/16	4 5/16	5	5 3/4	6 7/16	7 3/16	7 7/8
18	26	4	1/6	3/4	1 7/16	2 3/16	2 7/8	3 5/8	4 3/8	5 1/16	5 13/16	6 9/16	7 1/4	8
26	34	6	1/4	3/4	1 1/2	2 1/4	3	3 3/4	4 1/2	5 1/4	6	6 3/4	7 1/2	8 1/4
33	41	8	1/3	13/16	1 9/16	2 3/8	3 1/8	3 15/16	4 11/16	5 1/2	6 1/4	7 1/16	7 13/16	8 5/8
36	52	9	3/8	13/16	1 9/16	2 3/8	3 3/16	4	4 13/16	5 5/8	6 7/16	7 3/16	8	8 13/16
45	0	12	1/2	7/8	1 3/4	2 5/8	3 1/2	4 5/16	5 3/16	6 1/16	6 15/16	7 13/16	8 11/16	9 9/16
53	8	16	2/3	15/16	1 15/16	2 15/16	3 7/8	4 15/16	5 13/16	6 13/16	7 3/4	8 3/4	9 11/16	10 11/16
56	19	18	3/4	1	2 1/16	3 1/16	4 1/8	5 1/8	6 3/16	7 3/16	8 1/4	9 1/4	10 5/16	11 5/16

If span is in feet and inches, find result of feet span in left-hand table. Then find result of the inches of span in right-hand table and add two results together.

TABLE II (*contd.*)—LENGTH OF BACKED HIPS

Feet ← *Span of the Building* → *Inches*

Slope of Roof in Degrees		Rise of Roof in 12 in. (½ Span)	Pitch of Roof	20		21		22		23		24		1	2	3	4	5	6	7	8	9	10	11
°	′	in.		ft.	in.	ft.	in.	ft.	in.	ft.	in.	ft.	in.	in.	in.	in.	in.	in.	in.	in.	in.	in.	in.	in.
14	2	3	1/8	14	4 5/16	15	0 15/16	15	9 9/16	16	6 3/16	17	2 13/16	11/16	1 7/16	2 1/8	2 7/8	3 9/16	4 5/16	5	5 3/4	6 7/16	7 3/16	7 7/8
18	26	4	1/6	14	6 3/8	15	3 1/16	15	11 13/16	16	8 1/2	17	5 1/4	3/4	1 7/16	2 3/16	2 7/8	3 5/8	4 3/8	5 1/16	5 13/16	6 9/16	7 1/4	8
26	34	6	1/4	15	0	15	9	16	6	17	3	18	0	3/4	1 1/2	2 1/4	3	3 3/4	4 1/2	5 1/4	6	6 3/4	7 1/2	8 1/4
33	41	8	1/3	15	7 5/8	16	5	17	2 3/8	17	11 3/4	18	9 1/8	13/16	1 9/16	2 3/8	3 1/8	3 15/16	4 11/16	5 1/2	6 1/4	7 1/16	7 13/16	8 5/8
36	52	9	3/8	16	0 1/16	16	9 11/16	17	7 5/16	18	4 15/16	19	2 1/2	13/16	1 9/16	2 3/8	3 3/16	4	4 13/16	5 5/8	6 7/16	7 3/16	8	8 13/16
45	0	12	1/2	17	3 7/8	18	2 1/4	19	0 5/8	19	11	20	9 7/16	7/8	1 3/4	2 5/8	3 1/2	4 5/16	5 3/16	6 1/16	6 15/16	7 13/16	8 11/16	9 9/16
53	8	16	2/3	19	5 1/4	20	4 15/16	21	4 1/2	22	4 3/16	23	3 7/8	15/16	1 15/16	2 15/16	3 7/8	4 15/16	5 13/16	6 13/16	7 3/4	8 3/4	9 11/16	10 11/16
56	19	18	3/4	20	7 3/8	21	7 3/4	22	8 1/8	23	8 1/2	24	8 7/8	1	2 1/16	3 1/16	4 1/8	5 1/8	6 3/16	7 3/16	8 1/4	9 1/4	10 5/16	11 5/16

Slope of Roof in Degrees		Rise of Roof in 12 in. (½ Span)	Pitch of Roof	25		26		27		28		29		30		1	2	3	4	5	6	7	8	9	10	11
°	′	in.		ft.	in.	ft.	in.	ft.	in.	ft.	in.	ft.	in.	ft.	in.	in.	in.	in.	in.	in.	in.	in.	in.	in.	in.	in.
14	2	3	1/8	17	11 7/16	18	8 1/16	19	4 11/16	20	1 1/4	20	9 7/8	21	6 1/2	11/16	1 7/16	2 1/8	2 7/8	3 9/16	4 5/16	5	5 3/4	6 7/16	7 3/16	7 7/8
18	26	4	1/6	18	1 15/16	18	10 11/16	19	7 3/8	20	4 1/8	21	0 13/16	21	9 9/16	3/4	1 7/16	2 3/16	2 7/8	3 5/8	4 3/8	5 1/16	5 13/16	6 9/16	7 1/4	8
26	34	6	1/4	18	9	19	6	20	3	21	0	21	9	22	6	3/4	1 1/2	2 1/4	3	3 3/4	4 1/2	5 1/4	6	6 3/4	7 1/2	8 1/4
33	41	8	1/3	19	6 1/2	20	3 15/16	21	1 1/4	21	10 11/16	22	8 1/16	23	5 7/16	13/16	1 9/16	2 3/8	3 1/8	3 15/16	4 11/16	5 1/2	6 1/4	7 1/16	7 13/16	8 5/8
36	52	9	3/8	20	0 1/8	20	9 3/4	21	7 5/16	22	4 15/16	23	2 9/16	24	0 1/8	13/16	1 9/16	2 3/8	3 3/16	4	4 13/16	5 5/8	6 7/16	7 3/16	8	8 13/16
45	0	12	1/2	21	7 13/16	22	6 3/16	23	4 9/16	24	3	25	1 3/8	25	11 3/4	7/8	1 3/4	2 5/8	3 1/2	4 5/16	5 3/16	6 1/16	6 15/16	7 13/16	8 11/16	9 9/16
53	8	16	2/3	24	3 9/16	25	3 3/16	26	2 7/8	27	2 9/16	28	2 3/16	29	1 7/8	15/16	1 15/16	2 15/16	3 7/8	4 15/16	5 13/16	6 13/16	7 3/4	8 3/4	9 11/16	10 11/16
56	19	18	3/4	25	9 1/4	26	9 5/8	27	10	28	10 9/16	29	10 11/16	30	11 1/16	1	2 1/16	3 1/16	4 1/8	5 1/8	6 3/16	7 3/16	8 1/4	9 1/4	10 5/16	11 5/16

If span is in feet and inches, find result of feet span in left-hand table. Then find result of the inches of span in right-hand table and add two results together.

TABLE III—ROOF MEMBERS. CUTS IN 8 PITCHES

Slope of Roof in Degrees	*Rise of Roof in 12 in.*	*Pitch of Roof*	*Common Rafter Level Cut*	*Common Rafter Plumb Cut*	*Backed Hips Level Cut*	*Backed Hips Plumb Cut*	*Backed Hips Edge Cut*	*Backed Hips Backing*	*Jack Rafters Edge Cut*	*Purlins Side Cut*	*Purlins Edge Cut*	*Diminishing Length of Jack Rafter*
												ft. in.
14° 2′	3	$\frac{1}{8}$	14°	76°	10°	80°	45°	81°	44°	77°	46°	1 0$\frac{9}{16}$
18° 26′	4	$\frac{1}{6}$	18°	72°	13°	77°	45°	78°	43°	72°	47°	1 0$\frac{7}{8}$
26° 34′	6	$\frac{1}{4}$	27°	63°	19°	71°	44°	74°	42°	65°	48°	1 1$\frac{11}{16}$
33° 41′	8	$\frac{1}{3}$	34°	56°	25°	65°	42°	66°	39°	60°	49°	1 2$\frac{11}{16}$
36° 52′	9	$\frac{3}{8}$	37°	53°	28°	62°	41°	65°	38°	58°	51°	1 3$\frac{5}{16}$
45° 0′	12	$\frac{1}{2}$	45°	45°	35°	55°	39°	60°	35°	55°	55°	1 5$\frac{1}{4}$
53° 8′	16	$\frac{2}{3}$	53°	37°	43°	47°	36°	55°	31°	50°	58°	1 8$\frac{3}{8}$
56° 19′	18	$\frac{3}{4}$	36°	34°	47°	43°	34°	54°	29°	48°	61°	1 9$\frac{7}{8}$

Read downwards for different cuts of roof members and across for required pitch. All cuts to nearest degree.

Start at long point of plumb cut of common rafter. Mark the diminishing jack rafter length and repeat along rafter for pattern, giving lengths and number of jack rafters required at 14-in. centres.

TABLE FOR CONVERSION OF INCHES TO DECIMALS OF A FOOT

	Inches											
Inch	0	1	2	3	4	5	6	7	8	9	10	11
—	—	0·0833	0·1667	0·2500	0·3333	0·4167	0·5000	0·5833	0·6667	0·7500	0·8333	0·9167
$\frac{1}{8}$	0·0104	0·0938	0·1771	0·2604	0·3438	0·4271	0·5104	0·5938	0·6771	0·7604	0·8438	0·9271
$\frac{1}{4}$	0·0208	0·1042	0·1875	0·2708	0·3542	0·4375	0·5208	0·6042	0·6875	0·7708	0·8542	0·9375
$\frac{3}{8}$	0·0313	0·1146	0·1979	0·2813	0·3646	0·4479	0·5313	0·6146	0·6979	0·7813	0·8646	0·9479
$\frac{1}{2}$	0·0417	0·1250	0·2083	0·2917	0·3750	0·4583	0·5417	0·6250	0·7083	0·7917	0·8750	0·9583
$\frac{5}{8}$	0·0521	0·1354	0·2188	0·3021	0·3854	0·4688	0·5521	0·6354	0·7188	0·8021	0·8854	0·9688
$\frac{3}{4}$	0·0625	0·1458	0·2292	0·3125	0·3958	0·4792	0·5625	0·6458	0·7292	0·8125	0·8958	0·9792
$\frac{7}{8}$	0·0729	0·1563	0·2396	0·3229	0·4063	0·4896	0·5729	0·6563	0·7396	0·8229	0·9063	0·9896

SLOPE OF ROOF	COMMON RAFTER FEET	HIP RAFTER FEET
14° 2′	0·5153	0·7180
18° 26′	0·5270	0·7264
26° 34′	0·5590	0·7500
33° 41′	0·6009	0·7817
36° 52′	0·6250	0·8003
45°	0·7071	0·8660
53° 8′	0·8333	0·9718
56° 19′	0·9014	1·0307

THIS TABLE GIVES A CONSTANT WHIGH WHEN MULTIPLIED BY SPAN IN FEET GIVES LENGTHS OF RAFTERS IN FEET AND DECIMAL PARTS OF A FOOT. SHOULD ODD INCHES BE IN THE SPAN THESE CAN BE TAKEN FROM APPROPRIATE LINE IN MAIN TABLE AND ADDED TO THIS RESULT.

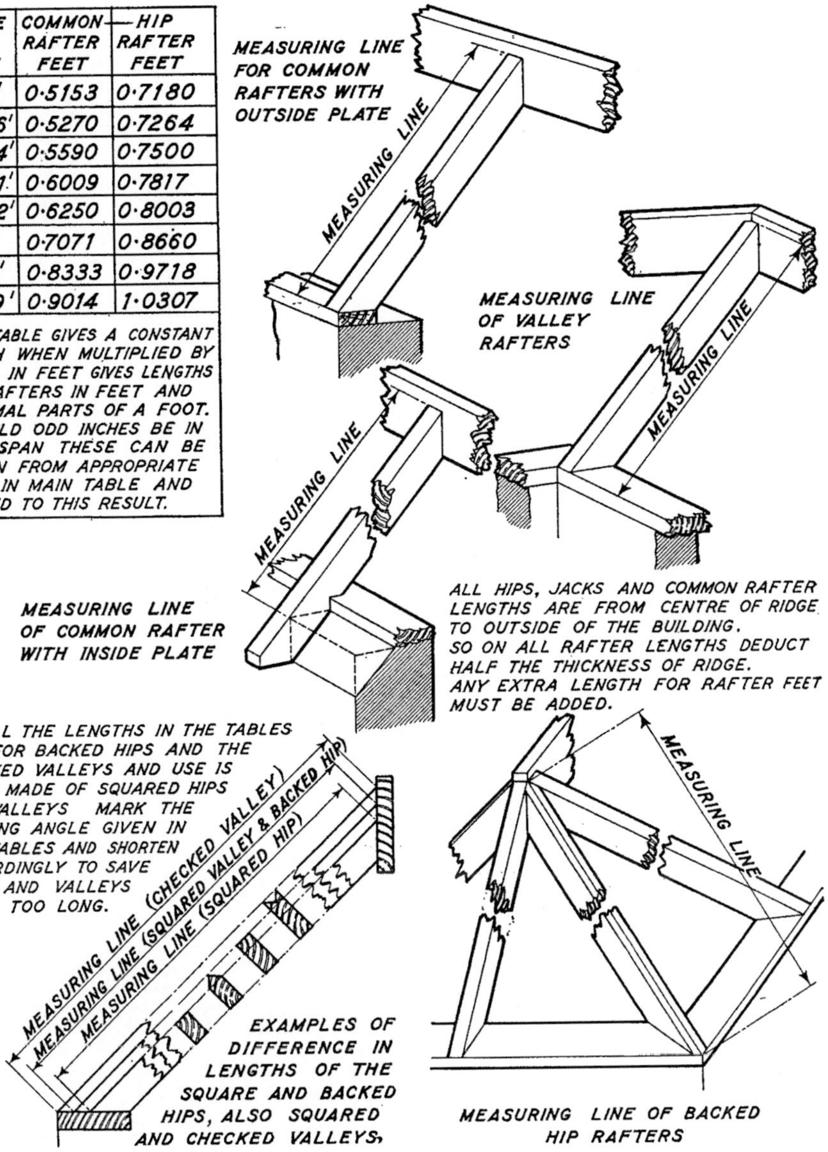

MEASURING LINES FOR ROOF MEMBERS

INDEX

A

Abutments in red cedar shingling, 112
Angle tie, 22
Asbestos cement, 96
Asbestos-cement corrugated sheets, 100
Asbestos-cement pantile, 100
Asbestos-felt, 107
Asphalte dressing to welted lead, 31
Asphalte roofing, 31
Asphalte shingles, 103

B

Battening, calculations, 63
Battens for plain tiling, 81
Beam :
 dragon, 22
 resistance moment, 45
 roof, 43
 straining, 27
Bedding materials, 102
Bitumen roofing felt, 114
Boarding, design of, 46
Bonnet hips, 77

C

Canadian red cedar shingles, 110
Cast copper nails, 83
Cedar-wood shingles, 111
Cement :
 bituminous, 109
 jointing, 108
 lap, for asbestos felt, 108
Clay pantiles, 100
Clear span, 1
Clout nails, 108, 123
Coachscrew, 32
Colour of wood shingling, 112
Common rafter, obtaining bevels and lengths, 54, 128, 129
Composite roof truss, 25, 27
Concrete roof abutting lantern light, 32
Corrugated-iron roofing, fitting, 116
Corrugated-iron sheets, 117
Covering gutters with ruberoid, 121
Covering valley with ruberoid, 121

D

Damp proofing of roofs, 36
Delabole slating, old, 59
Diagonal-pattern asbestos-cement slates, 96
Dormer window, 15, 16
Double-roll pantile, 102
Dragon beams, 22
Drip, 29

E

Eaves, the, 2, 123
 construction in slating, 65
 finish for pantile roof, 101
 finish for roofs, 14
 single-lap tiling, 89
 treatment of, 118
Effective span, 1
English oak shingles, 110

F

Fixing asbestos-cement corrugated sheets, 99
Flashing of ruberoid to brick walls, 121
Flashings, 106
Flat roofs, 28, 35
Flextile, 103
Flextile shingles, laying, 103
Fosalsil, 39
Fowler's formula, 43

G

Galvanised clout nails, 120
Galvanised hook bolts, 99
Galvanised iron screws, 99
Galvanised nails, 68, 82, 100
Gauge for plain tiling, 81
" Glass-crete," 33
Gutters, 36

H

" Handcraft " corrugated sheets, 98
Hand-made tile, 78
Hipped end, the, 2
Hips, the, 2, 101
 asbestos-felt for, 109
 covering of, in single-lap tiling, 91
 irons, 86
 rafter, obtaining bevels and lengths, 54, 130, 131
 single-lap tiling, 90
 wood shingles, 112
Hips and ridges, 105
Honeycomb-pattern asbestos-cement slate, 97

I

Imbrex, 72
Insulation, roof, thickness of material for, 36
Iron straps, 21

J

Junction of roof with lantern lights, 31

K

King-post roof truss, 18
frame diagram, 52
joints, 21

L

Lantern lights, junction of roof with, 31
Lath-and-plaster ceiling, 35

M

Machine-made clay tile, 78
Maximum span for rafters, 50
Modulus of rupture, 43
Mortar, hair, 102

N

Nailing, 82
cost, 83
wood shingles, 111
Nails, 82
cast copper, 83
clout, 108
copper, 83
galvanised, 82
heads, 109
yellow cast-composition, 83
zinc, 82

O

Old English pantile, 72
Open roofs, covering, 125
Overcloak, the, 30

P

Pantiles, 96
asbestos-cement, 100
double-roll, 102
fixing, 100
stop-end, 101
twin, 101
Parallel gutters, 36
Paropa, 38
Pitch of roof, 1
Plain tiling, 75
fittings, 76
mitred hip in, 87
rafter pitch for, 80
"Poilite" slates, 98
Purlins, 2, 8, 19, 44, 50, 55, 99

Q

Quarry tiles, 34
Queen-post roof truss, 23, 27

R

Rafters, 47, 50
jack, 2, 54
pitch for plain tiling, 80
working stresses, 43
Rankine-Gordon formula, 53
Repairing, 113
slate covering, 113
tile covering, 114
Ricker's formula, 43
Ridges, 98, 123
asbestos-felt for, 109
covering by bitumen roofing felt, 116
plain angle, for slating, 65
red cedar shingling, 111
single-lap tiling, 90
treatment, 120
wood shingles, 112
Ridges and hips, 105
Rolls for sheet metal, 29
Roofing:
asphalte, 31
copper, 30
materials, 44, 96
zinc, 30
Roofs:
bevels, 54, 127
collar-and-tie, 7
collar-tie, 4, 6
couple, 4, 6
couple-close, 4, 6
covering with bitumen roofing felt, 114
curb, 27
damp proofing of, 36
double lean-to, 9
double-rafter, 8
emergency repairs, 113
flat, 28
insulation, 35, 36
lean-to, 3, 5
Mansard, 24, 27
parts, 2
pitches, 1, 89
preparations, 81, 89
purlin, 8
single, 3
single Mansard, 11, 13
span, 1
trusses stress diagrams, 51
valleys, 9
Ruberoid covering, 121
Ruberoid roofing, fixing, 118

S

Sarking, asbestos-felt, 109
Screw roofing nails, 118
Shingles, asphalte, 103

Shingles, Flextile, laying, 103
Single Mansard roof, 11, 13
Slates :
  asbestos-cement, 96
  calculating number required, 59
  covering, repair, 113
  Delabole randoms, 63
  grading, 58
  holing, 64
  random, 63
  sized, 59
  sorting, 64
  straight-cover, 98
  weathering grey-green, 63
Slating :
  mitred hips for, 65
  plain angle ridge for, 65
  preparation, 64
  random, 69
  valley, 67
Spanish tiling, 74, 87
Sprocketed eaves for plain tiling, 82
Standard weight for asbestos-felt roofing material, 107
Steel construction of lantern light, 32
Steel sashes, 35
Stone slabs, 34
Stop-end pantiles, 101
Straining beam, 27
Straining cill, 27

T

Tapered hips, 71
Tapering gutter, 36, 41
Tegula, 72
" Thermotile," 39, 40
Tie-beam, 22, 27
Tiles :
  angle, 77
  Bridgwater double Roman, 70
  choosing, 78
  clay, 70
  cloister, 74
  concrete, 79
  Courtrai-du-Nord, 73, 89
  covering, repair, 114
  double Roman, 72
  double Roman interlocking, 73
  hip, 77
  improved interlocking pantile, 75
  Italian, 73
Tiles—*continued*
  Marseilles, 70
  old English pantile, 72
  pan, 96
  plain, 74
  Reynardo, 73
  ridge, 76
  single, 72
  Somerset interlocking No. 13, 73
  Spanish, 73
  special, 79
  " Thermotile," 40
  valley, 77
Tiling :
  bonnet hip, 86
  roofs, 70
  single-lap, 77, 89
  vertical, 88
Timbers, short, 10
Trussed roofs, 18
Twin pantile, 101

U

Undercloak, the, 29

V

Valleys :
  definition of, 2
  gutters, asbestos-felt, 108
  plain tiling, 87
  red cedar shingling, 111
  roof, 9
  single-lap tiling, 92
  tiling, 88
Verges, 2, 101
  plain tiling, 84
  red cedar shingling, 111
  single-lap tiling, 90
Vertical work in red cedar shingling, 111

W

Wall plates, 3
Walls in wood shingling, 112
" Watford " handcraft tiles, 98
Weights of plain tiles, 77
Weights of roofing materials, 44
Westmorland random slates, 62
Wood shingles, 110

Z

Zinc nails, 82
Zinc roofing, 30